T0325900

Food Safety and Quality Systems in Developing Countries

Food Safety and Quality Systems in Developing Countries

Volume Two: Case Studies of Effective Implementation

Edited by

André Gordon, PhD, CFS
**Technological Solutions Limited
Kingston, Jamaica**

AMSTERDAM • BOSTON • HEIDELBERG • LONDON
NEW YORK • OXFORD • PARIS • SAN DIEGO
SAN FRANCISCO • SINGAPORE • SYDNEY • TOKYO
Academic Press is an imprint of Elsevier

Academic Press is an imprint of Elsevier
125 London Wall, London EC2Y 5AS, United Kingdom
525 B Street, Suite 1800, San Diego, CA 92101-4495, United States
50 Hampshire Street, 5th Floor, Cambridge, MA 02139, United States
The Boulevard, Langford Lane, Kidlington, Oxford OX5 1GB, United Kingdom

Notices
Knowledge and best practice in this field are constantly changing. As new research and experience broaden
our understanding, changes in research methods, professional practices, or medical treatment may become
necessary.

Practitioners and researchers must always rely on their own experience and knowledge in evaluating and
using any information, methods, compounds, or experiments described herein. In using such information
or methods they should be mindful of their own safety and the safety of others, including parties for whom
they have a professional responsibility.

To the fullest extent of the law, neither the Publisher nor the authors, contributors, or editors, assume any
liability for any injury and/or damage to persons or property as a matter of products liability, negligence or
otherwise, or from any use or operation of any methods, products, instructions, or ideas contained in the
material herein.

Library of Congress Cataloging-in-Publication Data
A catalog record for this book is available from the Library of Congress

British Library Cataloguing-in-Publication Data
A catalogue record for this book is available from the British Library

ISBN: 978-0-12-801226-0

For information on all Academic Press publications
visit our website at https://www.elsevier.com/

Working together
to grow libraries in
developing countries

www.elsevier.com • www.bookaid.org

Publisher: Nikki Levy
Acquisition Editor: Patricia Osborn
Editorial Project Manager: Jackie Truesdell
Production Project Manager: Jason Mitchell
Designer: Matthew Limbert

Typeset by Thomson Digital

Dedication

This volume is dedicated to my late father, Justice Uel Dennison Gordon (Snr.) and my mother Cynthia Gordon who both encouraged me to make a positive contribution to my field, my region, and to my fellow man. Their guidance, faith, and love made this contribution to the field of Food Science and Technology possible. It is also dedicated to the many unnamed, and often unrecognized, food scientists and technologists working in the fields, the factories, the laboratories, and with consumers in developing countries under often trying circumstances, whose individual contributions continue to help to feed us with safe, wholesome food.

Contents

Contributors

A. Gordon
Technological Solutions Limited, Kingston, Jamaica

J. Jackson
Independent Consultant, Okemos, MI, United States

H. Kennedy
Technological Solutions Limited, Port of Spain, Trinidad and Tobago

R. Williams
Technological Solutions Limited, Kingston, Jamaica

Preface

This volume explores the keys to effective implementation of food safety and quality systems (FSQS), with a focus on selected, specific food safety and quality challenges in developing countries and how these can be mitigated. While the challenges and approaches examined are universal, the context for this series, this volume being no exception, is a developing country environment where human, technological, and financial resources are often constrained and the focus is typically on the competitive production for local consumption or export into third country markets. This set of circumstances calls for the application of approaches and technology grounded in theoretically sound food science and FSQS principles, with the flexibility and practicality born, in many cases, of necessity. What this does, therefore, is make the information presented here applicable in all circumstances in developed, emerging, and developing countries, particularly since the principles and approaches call on the same body of knowledge required by all food scientists and technologists, all FSQS practitioners and all market access professionals, wherever in the world they are located.

The first book in this series, among other things, examined in detail the cross-border collaborative approach that is sometimes required to solve thorny market access issues requiring the application of food science to address food safety issues for a traditional food containing a natural toxicant, *Blighia sapida* being the example. In this volume, we start by providing a general overview of some of the issues and considerations that impact effective implementation of FSQS and put this in the context of some of the more noteworthy foodborne illness incidents in the recent past. The unique and often very specific nature of the food safety or market access limitations associated with taking traditional foods from developing countries mainstream is illustrated with case studies examining national, industry-wide or firm-level issues, and the practical resolution of these. What is particularly unique about this body of work is that a variety of solutions are discussed to food safety and quality assurance issues through the presentation of detailed, specific approaches to real world challenges for a wide range of foods, from beverages and sauces to dairy products and canned vegetables. Challenges and solutions are illustrated throughout using case studies taken from various situations in which the authors were involved. These case studies are used to illustrate the successful application of sound food science principles and different methodologies to address food safety or quality problems and, critically, to the successful practical implementation of FSQS within a developing country context, in particular.

Readers will encounter in this book a rich source of previously unpublished information about the practical application of food science and technology to solve food safety and quality problems in the food industry. They will also find it an invaluable resource as they seek to implement FSQS in companies or solve seemingly intractable and costly problems with products in the categories covered. Finally, this book will provide a treasure trove of information on tropical foods and their production that have applicability to similar foods and facilities around the world. We expect students, researchers, FSQS professionals, regulators, and market access practitioners to find this book an irreplaceable addition to their arsenal as they deal with issues regarding food safety and quality for the products with which they are working.

André Gordon, PhD, CFS

Acknowledgments

The editor and his coauthors acknowledge the dedication to duty and support provided by the Elsevier team on this project, particularly Ms. Jaclyn Truesdell and Ms. Karen Miller, both Editorial Project Managers, and Ms. Patricia Obsourne, Senior Acquisitions Editor (Book Division) who have helped, encouraged, and guided as the manuscript went through its phases. We also wish to thank the support team at Technological Solutions Limited (TSL) who undertook key aspects of the research, information gathering and provided the kind of on-going support that helped to make this possible. We express our deepest gratitude to Mr. Andrew Ho, GraceKennedy & Co. Ltd., Mr. Sean Garbutt, AML/Walkerswood Caribbean Foods Limited, Mr. Uriah Kelly, UL Manufacturing Limited, Ms. Simone Hew, Fachoy Foods Limited, Mr. Andrew Gray, Gray's Pepper Products Limited (Jamaica), Ms. Merlissa Khan, RHS Marketing Limited (Trinidad and Tobago), and Mr. Kenneth Da Silva (St. Vincent and the Grenadines). The authors and, we believe, you the readers owe a debt of gratitude to these persons and their companies that so graciously agreed to have their cases shared so that others could benefit from their experiences. Had they not allowed me to share their stories, this book would not have been enriched as it has by the information derived from those case studies. Finally, I thank my family for continuing to share my time to allow me to undertake and complete this exciting project.

Introduction: effective implementation of food safety and quality systems: prerequisites and other considerations

1

A. Gordon
Technological Solutions Limited, Kingston, Jamaica

Chapter Outline

Introduction

The food industry in developing countries is as diverse as its peoples. Depending on the country and the cuisine in that country, the food industry covers a wide range of sectors from seafood, fresh produce and traditional foods, fruits and vegetables, meats (processed and unprocessed), dairy products, beverages, confectionary, and grains and cereals among others. The characteristics of the food and the nature of the industry in each category of food varies by country and by region. It depends on the history, climate and microclimate, cultural and agricultural practices, the existence of a significant formal hospitality industry, the existence and level of development of a formal food service sector [including domestically grown or transnational quick serve restaurants (QSRs)], and the range and nature of the exports markets with which trading relationships have been established. These, collectively determine the range, breadth, and sophistication of the food industry in each region of the developing world and each country. For countries which share a border, the differences in the industry across a distance of only a few miles, indeed a few meters, can be quite sharp. Examples of this in different regions include The United States of America and Mexico in the Americas, India and China in Asia, Haiti and the Dominican Republic in the

Food Safety and Quality Systems in Developing Countries. http://dx.doi.org/10.1016/B978-0-12-801226-0.00001-3

Figure 1.1 Indonesian satay.
Source: www.indostyles.com.

Caribbean, Botswana and South Africa in Africa, and Suriname and Brazil in South America.

Typically, the foods consumed in developing countries include a wide and delicious range of produce, meat, poultry and seafood products, grains, fruits and vegetables, many of incomparable taste and flavor and many of which may be unknown in, or unfamiliar to, developed country markets. These include products such as *fungi* (pronounced *funji*, made with cornmeal) from the Eastern Caribbean, *cassreep* from Guyana, *dukunoo* (pronounced *dokoonou*) which originated in West Africa and is found throughout the Caribbean, *tamales*, *empanada* and *yuca* which are very popular in Central America. They also include *satay* (meat on a stick roasted over an open fire or charcoal) which is a staple throughout Southeast Asia and which is now gaining acceptance in some metropolitan markets, particularly some European markets (Fig. 1.1) and, from Africa, *gari* and *fufu* (Western Africa) and *fried plantains* (popular in the continent, in general). From these cultures have also come a range of new meal options and products which are sought after in developed country cuisines (International Food Information Council (IFIC) Foundation, 2013) including *curries* from Asia, *jerk* from Jamaica in the Caribbean (Fig. 1.2), *tacos* and *burritos* from Mexico in Central America, *falafel* from Lebanon in the Middle East, and *chow mien* from China. These are buttressed by equally delicious processed foods, the variety of which vary depending on cultural nuances and the sophistication of the food industry in the country (Fig. 1.3). Developing countries are also major producers of well-known and loved foods, widely consumed in developed countries, such as cocoa and cashew nuts (Ivory Coast), coffee (Brazil and Columbia), conch—*Strombus giga* (Jamaica), pineapples (Philippines and Costa Rica) and kiwi fruit from Chile.

Many developing countries produce a variety of foods which include fruits, vegetables, meat, seafood and dairy products, as well as processed foods and beverages

Figure 1.2 Jamaican jerk chicken.
Source: Walkerswood Caribbean Foods.

(Fig. 1.4) of a wide range of types, packaging, and formats. These are served through delivery channels as diverse as you will find in the more developed countries in Australasia, Southern Africa, the Middle East, Europe, and North America. Many of these are produced by staff trained and equipped to make high quality products. They are produced in facilities ranging from very basic to state-of-the-art (Fig. 1.5). The role and importance of food safety and quality systems (FSQS) in the food industry of developing countries and emerging economies, therefore, will depend on the level of sophistication and focus of their domestic industry and, specifically, the subsector of the industry from which the product comes.

Many traditional products are already being exported to markets globally. Traditional ethnic cuisines are now increasingly an inseparable part of mainstream diets

Figure 1.3 Developing country foods with different levels of sophistication.
Source: A. Gordon, 2015.

Figure 1.4 Examples of the some foods produced in developing countries.
Source: A. Gordon, 2015.

throughout the world. Chinese cuisine, Indian curries, Thai food, and traditional Asian cuisine are now not only available in specialty restaurants, but are a standard part of meal choices in Europe and North America. Many of these are available as authentic meal options in convenient formats in the major multiples[a] such as Whole Foods (Fig. 1.6A) and Walmart in the USA, Sainsbury and Tesco (United Kingdom), Loblaws (Canada), and Albert Heijn in the Netherlands (Fig. 1.6B). In addition to these prepared meals, many of the other foods and beverages consumed in the South have been adopted by the more developed country markets. These traditional developing country foods now included in global diets have expanded to include a wide

Figure 1.5 Examples of different kinds of food processing systems in developing countries.
Source: A. Gordon, 2015.

range of fruits and vegetables, produce, spices, prepared foods, and beverages. Most of the produce that form a part of this trend usually undergo further processing after purchase, including various forms of preparation and cooking.

In the initial stages when foods from developing countries started being sold in metropolitan marketplaces, they were originally sold mainly to the ethnic community familiar with them. Many have transitioned and are now being sold to mainstream populations and are therefore present in several conventional retail outlets alongside

Figure 1.6 Leading products from developing countries on sale in Whole Foods (A) and Albert Heijn (B).
Source: A. Gordon, 2015.

the prepared foods mentioned previously. Some of these products have been used by consumers around the world for decades by persons who perhaps even today remain unaware of their origin, examples being Angostura Bitters and Pickapeppa Sauce (Fig. 1.7A–B, respectively). Some are global standard bearers in their food class, being regarded as sophisticated world brands such as Guyana's well known El Dorado 15 Year Old Rum, one of the standard bearers in the rum category of the alcoholic beverage industry, or Blue Mountain coffee (Fig. 1.7C–D). Examples of some of the products that are now a part of global diets which originate almost exclusively from developing countries, as well as cuisines are summarized in Table 1.1. In essence then, the nature of our global food supply is changing to become much more diverse and this will bring new opportunities, even as new challenges will also emerge.

Many developing country exports are traditional foods that have continued to maintain old traditions and practices. These sometimes present interesting challenges for validation of their safety in a manner that meets the requirements of developed country market regulators. A prime example of this is the ackee (Fig. 1.8), Jamaica's National Fruit which was discussed in some detail in Volume I of this series (Gordon, 2015a). The aspect of the food industry that produces mainly for local consumption, although not export oriented as other aspects of emerging market food industries tend to be, is no less diverse as they face, in many cases, relentless pressure from domestic and imported competition. Depending on the country, the sector may produce a wide variety of products ranging from the relatively basic to sophisticated. Regardless of the nature of the food being imported or produced in different countries, the factors affecting the future development of the sector will be the same as they are elsewhere. It is important, therefore for both the producers and exporters, as well as importers in developed countries, to become aware of the international trends that are driving the development of the global food industry and equip themselves to meet the demands as they seek to have their products become leaders in the global marketplace. We will examine these throughout this book, starting with this chapter, and discuss how they relate to the overwhelming and increasing influence of FSQS in global food trade.

Figure 1.7 Examples of standard bearers and market leading products from developing countries on sale in the major multiples.

Figure 1.8 Jamaican ackee on the tree.

Food safety and quality

Ensuring safe food

All industry stakeholders in developing countries have a crucial role to play in ensuring that the products being manufactured or produced and sold in their countries or exported meet basic food safety standards. Within these countries, customers buying the food expect good quality, healthy food; the local regulatory infrastructure has a primary responsibility to assure this. For safe, good quality food, consumers will pay a fair price. It is therefore in the interest of producers and food industry professionals supporting them to ensure that the food sold to customers, wherever located, meets and/or exceeds their expectations. The food ideally should be good for them, healthy and well presented, delivered in a manner that allows them to trust that the purveyors are delivering on the promise of the entire value chain. What is that unspoken promise? That all food being delivered, offered for sale, or served is *safe*, at the very minimum. The expectation, the promise is that the food being consumed will not make customers ill. Therefore, the system of production, from land identification, through planting, handling, processing, and delivery to the market in developing or developed countries has to ensure that the increasingly rigid requirements of the market and expectations of consumers are met.

Ensuring the safety of food means preventing the entry of hazards into the food being handled or eliminating or controlling them, where they are naturally present in the food. These include physical, chemical, and biological hazards, including microorganisms (Fig. 1.9). In Volume I of this series, the latter was explored in some detail for a range of foods, particularly with regard to chemical hazards (Gordon, 2015a). In cases where the products are being sold in their market of origin, locals, visitors, the suppliers of visitors to markets (including big hospitality businesses such as the TUI Group[b]), hotels, and their supporting infrastructure also expect absolute safety of the food being offered. For the latter this is of great importance since they often face significant liability should a visitor or a customer to which a vacation was sold

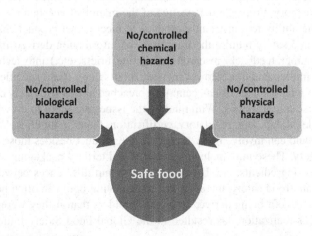

Figure 1.9 Delivering safe food.

become ill (Kitchin, 2015). The cost of the litigation which can arise from any such situation can be quite substantial (Cooper, 2014) and, in some cases depending on the situation, could cause the business to fold, examples of which will be examined later in the chapter. Consequently, purveyors of foods now have to be mindful not only to focus on the demand for quality products, delivering acceptable nutrition, but also on the safety of the food products being offered for sale, consumption, or import. This latter has largely been driven by the tidal wave of concern about food safety that has become even more important over the last decade or so and is now a major driver of the future direction of the global food industry as is explored here.

Safety versus quality

The management of food safety and quality, while different, are interrelated. Food industry practitioners will find that in seeking to ensure the safety of the food they are handling or producing, they will be able to achieve the delivery of a quality food as well. It is, however, quite important to carefully and clearly distinguish between those aspects of a food and its production or handling that are safety related, related to quality, or both. The reason for this is that misunderstandings between what comprises food safety and quality have led to many of the egregious errors in food handling that have caused unfortunate illnesses and, in some cases, avoidable deaths. Safety is absolute and unforgiving. A food is either safe or unsafe; there is no in-between. It is typically a prerequisite that is demanded by countries through compliance with their requirements to access their market. Increasingly, developed countries are mandating that foods entering their countries, as is the case for foods produced in their own jurisdictions, must meet minimum food safety standards, including GMP and/or GAP, HACCP, traceability, and other requirements.

Quality, on the other hand, is primarily focused on meeting (and exceeding) the customers' requirements and seeking to delight the customer, who *already assumes* that the product being offered *is safe*. Product quality is also related to entry into selected market segments by meeting specific quality criteria set by buyers, retaining and expanding market share, and commanding a higher premium within a particular product category. From the perspective of the manufacturer, quality may be seen as assuring the ability to consistently produce to meet preset product characteristics. Both quality and safety require the translation of information derived from multiple sources (consumer feedback, research, marketing briefs, etc.) into technical specifications for implementation. However, with food safety, these specifications often become limits which, once critical, cannot be breached as this will likely cause illness, injury, or worse to consumers. With quality, the issues are more to do with noncompliance with legal, labeling, regulatory, certification, buyer, retailer, and/or consumer requirements and can involve many aspects of the product besides those that directly impact its safety. These may include secondary and tertiary packaging, color, flavor, labeling, sizes, ingredients, etc. Despite the important differences between ensuring food quality and food safety, the approach to achieving both can often be combined, with a greater benefit being derived to all stakeholders than if they were approached separately. This realization has resulted in the Global Food Safety Initiative (GFSI) benchmarked schemes all being combinations of food quality and food safety systems (Global Food Safety Initiative, 2016a,b), as well as a trend in which some of the stipulations of regulators are seeking to incorporate many of the elements that are a part of the GFSI set of benchmarked standards. The GFSI and its set of standards is discussed in more details in subsequent sections.

This series of volumes and this volume in particular, including many of the case studies outlined throughout, have relied on the principle that focusing on food safety, done well, will address most of the critical issues to do with quality and that where this does not hold true, the quality issues can be addressed with the same approach or framework. This approach involves identifying those areas of the production and handling process where control must be exercised to deliver the desired outcome, whether it is taste and flavor, color, choice of packaging, compliance with requirements, or other important quality requirements. Throughout, case studies and evaluation of situations that have arisen are presented in a manner that allows the reader to understand the issues, see how these are defined in technical terms and then addressed through the systematic application of carefully developed approaches to achieving a desired food safety or quality goal.

Food safety drivers

Food safety and the assurance of food safety have become among the major concerns of consumers in developed country markets over the last several years (Global Food Forums, 2016). No longer a "nice to have," food safety is now an absolute requirement, now (virtually) mandatory as it is the price of entry into locally based, international hotel chains in developing countries and all major export markets. For companies wishing to grow their businesses by expanding in the local market and gaining competitive leverage, being able to assure the safety of their food products and communicate this effectively to consumers is a distinct advantage. Foodborne illness outbreaks have led to calls from consumers for governments to ensure a safe, secure food supply. They have also led to increasing pressure on food businesses (retailers, manufacturers, etc.) to ensure compliance. Foodborne illness outbreaks have also driven all major markets to act, including the USA, Canada, and the United Kingdom/European Union, as we shall examine throughout the course of this book. Hence the changes in EU regulations over the years, the tightening import regulations into the ANZAC region from neighboring countries in Asia, the Food Safety Modernization Act 2011 (FSMA) in the United States, the Safe Food for Canadians Act (SFCA) 2012, and the Global Food Safety Initiative (GFSI).

What has driven this continuing thrust?

The fact is that the number, spread, and impact of foodborne illness outbreaks and related incidents have increased significantly over the last decade and consumers and consumer protection groups have become ever more vigilant, even as governments and the producers themselves have become more demanding of the entire industry value chain. The impact of foodborne illness outbreaks as illustrated in multiple countries from several continents in Table 1.1 demonstrates the significant economic and social impact of foodborne disease. As in the United States and other developed

Table 1.1 **Impact of foodborne illnesses in selected developed countries 2015**

Country	Number of cases of illness	Hospitalizations	Deaths	Estimated cost (US$)	Population
United Kingdom[a]	1 million	20,000	500	1.99 billion	65,000,000
United States of America	48 million	128,000	3,000	77.7 billion	323,644,000
Australia	4.1 million[b]	—[c]	120	0.97 billion	24,277,000
Canada[d]	4 million	11,600	238	1.1 billion[e]	36,242,000

[a]Food Standards Agency (2016).
[b]New South Wales Government, Department of Primary Industries Food Authority (2016)
[c]Data not found
[d]Bélanger et al. (2015).
[e]Todd (1989).

countries, in Australia, a leader in the Pacific region, efforts by the authorities have resulted in a decline in foodborne illnesses from 5.4 million in 2011 to 4.1 million by 2015 (New South Wales Government, 2016), a figure which is nevertheless too high. When reduced to the national, regional, or local level, the impact is even greater and, when vulnerable populations which comprise nearly 20% of national populations in major developed countries (Lund, 2015) are taken into consideration, the consequences are even more dire.

In a well-publicized case in Australia, then 7-year-old Monika Samaan was left brain damaged for life from salmonellosis reportedly resulting from eating a popular meal at a QSR that was contaminated with *Salmonella*. While the QSR maintained that the evidence was not clear that their food was the cause of the illness, Monika's family was awarded $8.08 million in damages. This incident, which started when the food poisoning occurred in October 2005, followed other well publicized cases, include the watershed food poisoning outbreak from drinking water in Walkerton, Ontario 5 years earlier in 2000 (Clark et al., 2003). Several other notable cases that will be explored later, have all combined to make it an absolute imperative that government and industry ensure that the food being delivered to consumers is safe.

In many parts of the developing world, the kind of detailed data on the impact of foodborne illnesses is not gathered in a manner that allows it to be credible and, as such, the information needed to plan effective mitigation strategies is not available. In India, the data gathering to properly advise national foodborne illness prevention programs and strategies also suffers from this weakness (Rao et al., 2012), although data available for 2004 showed that just over 9.57 million cases of gastrointestinal illnesses caused by food- or waterborne vectors was reported (Sudershan et al., 2014). In the Latin American and the Caribbean, for example, although limited in occurrence, there have been significant domestic cases of foodborne illnesses (Pires et al., 2012), some of which have caused great dislocation in the local communities (Indar, 2012). Consequently, regardless of the availability of corroborating figures in developing countries, it is clear that the issue of food safety and quality also needs informed and constant attention.

Food safety and quality considerations in accessing and developing export markets

As has been discussed, many producers and exporters from developing countries, as well as importers in developed country markets, are moving to capitalize on a growing demand, particularly among the millennials (Sanders, 2015) for a wider variety of food choices, including what used to be regarded as "traditional" foods from developing countries. With products such as *sriracha, various kinds of rice, tikka masala*, and *harissa*[c] (Fig. 1.10) becoming hot items in the mainstream market (Sanders, 2015) and the continuing trend toward tastes for new experiences (Institute of Food Technologists, 2015), food safety remains among the most important drivers of what will be acceptable in the marketplace. This has taken the form of a range of considerations which developing country food industry stakeholders need to be mindful of and develop strategies and effective approaches to deal with if they intend to capitalize on the favorable trends. Depending on the market for which the food item is being produced, the following are among these considerations:

- Local Regulations (in the target market) and their enforcement
- Food Safety
  - Hazard Analysis Critical Control Points (HACCP)
  - Hazard Analysis and Risk-Based Preventive Controls (The Preventive Controls Rule-21 CFR 117)
  - The Produce Rule (21 CFR 112)
  - The Safe Food for Canadians Act (2012)
  - The EU regulations governing the particular food item that is being produced and exported.
- US Public Law 107-188 (Food Safety and Bioterrorism Act)
- The FDA Food Safety Modernization Act (FSMA)
- Regulatory differences between Canada, European Union and United States if the item is targeted at multiple markets
- Traceability and Recall requirements
- Good Agricultural Practices (GAPs)
- Good Manufacturing Practices (GMPs)
- Packaging and Labeling requirements
- Residues (Pesticides, Heavy Metals, Others)
- Different Certification Requirements (market driven)
- Global Food Safety Initiative (GFSI) Standards

All of these have to be considered and effective strategies sought to deal with them, as relevant, depending on the market and specific market channel into which the country or specific producer is exporting.

Figure 1.10 Types of rice, harissa, and tikka masala on retail shelves.
Source: A. Gordon, 2016.

Basic requirements of a food safety system

There are several aspects of a food handling or production operation that need to be considered and effectively addressed if a comprehensive, robust FSQS is to be established and maintained. Among these requirements the following deserve particular focus:

- the physical premises;
- plant or facility layout and design;
- product and personnel flows throughout the facility;
- training, staff competencies, and competence determination;
- traceability throughout the entire handling chain
- recall program;
- sampling and testing;
- inventory control;
- pest control;
- standard operating procedures (SOPs);
- sanitation and sanitation SOP (SSOPs);
- labeling and coding;
- internal and external auditing;
- storage and transportation;
- preventative maintenance; and
- supplier quality assurance (SQA).

The organization, housekeeping, and security of the physical *premises* are important as this is the first point of control of the manufacturing/food handling environment. If the premises are below standard or security is compromised, it will be difficult to ensure food safety and implement an effective food defense program. The *plant layout and design* determine the overall efficiency of the plant and can significantly impact food safety as, among other things, this often determines *product and personnel flows* throughout the facility. It also influences the efficacy of the *storage and transportation, inventory control* and the ease and effectiveness of *pest control* and maintenance of the facility. As a part of ensuring the overall safety and quality of the food being handled in the facility, *pest control, sanitation* governed by *sanitation standard operating procedures*, effective *preventive maintenance*, and the *training* of all staff to ensure their competence are critical. *Staff competence* to undertake each specific task that impacts food safety, quality, or the efficiency, and therefore the profitability of production must be demonstrated by objective verification through a documented system of *competence determination* for which adequate records are kept. This will require a comprehensive training program, supported by the appropriate documentation and records of the training done.

An important aspect of any modern food handling system is *traceability* throughout the entire handling chain from the sourcing of each raw material and input to the delivery to the market (the final customer of the firm). This is critical to support an effective *recall program* that must be able to track, trace, and support the recall of any finished product within 2 h (now the standard time for most private certification systems globally). This requires an effective labeling and coding system that is compliant with all relevant regulations in the products intended markets, and is the foundation for world class inventory control and warehousing as part of the storage and distribution program. The control of the all inputs into the production process, including packaging, is best managed by carefully crafted *SQA programs*, inclusive of suppliers' Letters of Guarantee (LOG), where relevant, and certificates of analysis (COAs) for raw materials supplied. Both of these may be essential to assure the firm that the input or raw material being supplied meets the *specifications* for each input which would have either been developed for them or are required to meet customer or regulatory requirements. SQA programs also include statistically valid *sampling and testing* programs and both self-assessment through internal audits, as well as the periodic assessment of suppliers. The sampling and testing program typically cover verification of the quality and safety of the products made or handled as part of the food safety and quality assurance programs employed, inclusive of assuring full compliance with all *pesticide, chemical, heavy metal, antibiotic*, and *hormonal residues*, as relevant and as required.

The programs discussed previously are among, or a part of, required *prerequisite programs (PRPs)* that provide the support needed for the HACCP, as well as the quality assurance components of the FSQS to be able to operate effectively. These PRPs control the general conditions in the food handling or production environment that assure overall safety and compliance of the food being handled with regulatory, industry, and customer norms and requirements. Other programs not previously mentioned include *GMPs, Cleaning and Housekeeping, Consumer Complaints Management, Allergen Management*, and an *Analytical (Testing) Program*, which would include defined sampling plans and protocols. All of these programs must be structured in a manner that allows them to be verified and validated, as well as easily introduced to new team members as part of their orientation and training program. All programs must also be documented in such a manner as to unambiguously describe all aspects of the program, particularly how things are to be done, as typified in standard operating procedures (SOPs). How a typical sanitation PRP should be structured is exemplified by the outline of an effective sanitation program discussed in subsequent sections.

Example of the structure of an effective PRP: sanitation programs

A compliant, effective sanitation program must be well structured and have a detailed description of what is to be done, when it is to be done, how it is to be done, and by whom, captured in *Sanitation Standard Operating Procedures (SSOPs)*. The program should be documented such that all aspects are clear and easily explained in unambiguous language to the persons responsible for executing various aspects of the program. The program will involve housekeeping, cleaning, and sanitization which all work together to keep all areas of the plant and its surroundings in a condition conducive to the production of high quality, safe, regulatory compliant food products. Housekeeping is the control and removal of dust, clutter, extraneous material and ensuring order inside and outside of the facility and its environs. Cleaning is the removal of dirt and filth through the application of mechanical, physical, or chemical means, including the use of detergents and other cleaning agents. Sanitization is the application of specific sanitizers in

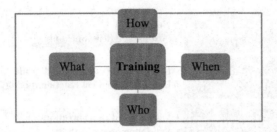

Figure 1.11 Overview of the design of a training program for sanitation.

a manner and at a concentration such that they will effectively, alone or in combination, reduce to an acceptable level, eliminate or control the microorganisms being targeted. When combined and executed effectively, these will form a cleaning and sanitation program that will deliver on the objective of a high-quality, low microbial load environment that is unlikely to cause contamination of the products being handled.

All PRPs must be thoroughly documented. As such, not only the records used, but a description of their use and the purposes for which they are used must be clear. The program must clearly indicated, for each surface, area, piece of equipment, or room being cleaned and/or sanitized:

- the *frequency* of cleaning and sanitation;
- the *responsibility* for each task or aspect of the program;
- the *specific piece of equipment or area of the facility* being cleaned;
- the specific *detergents, cleaning agents* and any *other chemical or sanitizer* to be used;
- the *specific concentrations* of each soap, cleaning agent or sanitizer used;
- the details of *how* the activity is to be done; and
- *how will effectiveness be verified.*

Zoning is often important as those areas which are designated "high care" areas require the sanitation and cleaning program to be effective in eliminating potential contaminant organisms while in "medium care" or "low care" areas reduction to a minimum level or control of their presence and growth might be acceptable. Whatever the overall design, the program will need to be supported by a training program that ensures the competence of all of the staff involved in housekeeping, cleaning, and sanitation. As depicted in Fig. 1.11, this program will have to describe *how* the housekeeping, cleaning, or sanitation activity is to be done, *what* is to be done, *who* is to do it, and *when* it is to be done. Otherwise different people may undertake the same task and obtain differing results. Training also should cover how the entire program is to be documented and the records to be used, as well as how they are to be managed.

Other considerations

Sanitation programs (inclusive of the housekeeping and cleaning components) also have specific requirements for the chemicals being used, in addition to those already mentioned. They require that all chemicals being used are only handled by trained and competent (and verified to be so, this also being documented) personnel. The storage of these chemicals must be secured and both the storage and handling must be by specifically

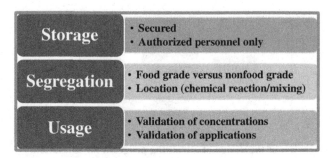

Storage	• Secured • Authorized personnel only
Segregation	• Food grade versus nonfood grade • Location (chemical reaction/mixing)
Usage	• Validation of concentrations • Validation of applications

Figure 1.12 Critical aspects of the management of a sanitation program.

authorized personnel (Fig. 1.12). All chemicals being used throughout the facility for the cleaning and sanitation of food-contact surfaces should be certified safe for use in a food handling facility (i.e., certified food grade). Nonfood grade chemicals may be used to clean and sanitize nonfood contact surfaces, such as floors, walls, and the exterior of the facility. For all chemicals used in the operations (for whatever purpose), it is required that their concentrations and also the mode and specifics of how they are applied are validated. As such, the amounts of each chemical used, the specific concentration at which it is used, how it is mixed (supported by documented mixing instructions), how its concentration is determined, and how it is applied must be recorded. Best practice also suggests that verification procedures are developed, documented, and applied at a predetermined frequency for all aspects of the program and includes validation of the suitability, security, concentration, and application of the chemicals used (Fig. 1.12).

Sanitation tips

For sanitation programs, it is expected that all chemicals and their concentrations of use are documented, that there is a comprehensive list of approved chemicals on file, as well as copies of all labels and the material safety data sheets (MSDS) for each chemical. The validation of concentrations and application as well as verification that there is appropriate management of storage and segregation of chemicals is also required. As for all aspects of FSQS, the well-known adage applies "If it is not documented, it doesn't exist or didn't happen!" Consequently, records of all activities as well as appropriate SSOPs are required.

Summary of PRPs

Effective PRPs are fully documented and applied in accordance with the documentation. They are validated as being effective, staff are trained to properly implement and manage them and records are kept used, in part, to verify that they are being implemented and maintained as intended. When developed, implemented and documented as indicated previously for sanitation programs, a comprehensive set of PRPs such as described in this section form the strongest possible support platform for effective FSQS that are fully compliant with both regulatory and specific market requirements. These are further explored throughout this volume and the case studies discussed in various chapters.

Conclusions

This chapter examined the wide diversity of food products from developing countries and their growing influence on the diet of consumers in developed countries and around the world. The importance of food safety and quality was discussed. While food safety and quality are inextricably linked, production of a quality product depends on the product first being safe. This is influenced by a wide variety of considerations, including the nature of postharvest handling of the product or its ingredients and food safety systems employed. In this regard, a HACCP-based FSQS, supported by a comprehensive set of PRPs, can deliver a program that is not only compliant with all relevant regulations in the markets of interest, but also can consistently provide safe, high quality food. Further, if the PRPs are organized, implemented, and documented in a systematic manner, then the overall operation will facilitate the market access required to support the expansion of businesses and diversification of the global food supply. In the ensuing chapters, practical examples of dealing with food safety and quality challenges as well as market access issues are explored. These are presented as case studies that are intended to guide producers, exporters, and FSQS practitioners in their quest to ensure the delivery of safe, quality food to consumers across the world.

Endnotes

[a]Multiples are major retailers that have many stores across several developed countries and exert significant influence on the market and market trends.
[b]TUI Group is the world's largest travel business, operating six airlines with a fleet of over 130 aircraft, 13 cruise ships and over 300 hotels worldwide. TUI serves more than 30 million customers.
[c]Sriracha is a Thai dipping sauce product made from a base of a hot chili pepper paste. Tikka masala is a spicy Indian sauce, and harissa is a traditional Maghrebian hot chili pepper paste, popular in the North African countries of Tunisia, Algeria, Libya, and Morocco.

Food safety–based strategies for addressing trade and market access issues

2

A. *Gordon*
Technological Solutions Limited, Kingston, Jamaica

Introduction

The foods produced and traditionally consumed in developing countries, including their fruits, vegetables, beverages, sauces, and entreés are increasingly being adopted by consumers in developed countries as a part of their diet. This is in keeping with a trend toward more healthy traditional foods and diversification of diets with exotic foods [International Food Information Council (IFIC) Foundation, 2013; Sloan, 2014]. This trend has accelerated over the last 10 years such that the percentage and range of food imports into developed country markets from developing countries continues to grow each year. This has meant that all markets worldwide are seeing a convergence of the requirements for food handled and sold and nowhere is this more visible than in the area of food safety and quality. Many of the traditional importers of foods from developing countries have been doing so for decades, with their products targeted mainly at the ethnic (diaspora) market. Due to this, the insistence on food safety and quality systems (FSQS) had not been a major issue for these importers and, consequently, neither was it a major consideration for the corresponding exporters in the source developing countries. With the increasing trend toward the mainstreaming of tropical and developing country foods, while many of the original importers remain in the business, the entry of new importers and buyers, including many of the major multiple themselves, has completely changed the dynamic. Another related and important change has been the opening of new distribution and marketing channels that previously had not been a part of the distribution chain, to serve many of the new

Food Safety and Quality Systems in Developing Countries. http://dx.doi.org/10.1016/B978-0-12-801226-0.00002-5

Figure 2.1 Range of developing country foods available on the shelves in an outlet of a developed market retail outlet.
Source: André Gordon, 2016.

consumers. This has been accompanied, to some extent, by the shifting of traditional consumers of developing country imports over to these new distribution channels, as the products they once had to travel to specific locations in major cities to get are now becoming available in many neighborhoods, in major retail outlets (Fig. 2.1), convenience stores, and even pharmacies in the major metropolises.

What these changes have meant is that products such as pineapples from Costa Rica, Thailand, and Cote d'Ivoire long available in retail outlets, are now being sold alongside mangoes (*Mangifera indica*), traditionally sourced from Mexico, India, and the Philippines. These have been joined by mangoes from Haiti who over the period 2002–06, with assistance from the USAID's[a] Hillside Agricultural Project Program (HAP), had firms that not only upgraded their hot water treatment facilities to eliminate the fruit fly from mangoes destined for the United States, but also implemented GMPs and HACCP-based food safety programs to comply with changing import standards. The changes have also meant that exporters of processed foods such as South Africa, the Philippines, Mexico, Chile, Thailand, and Brazil, among others, have had to implement or upgrade their food safety systems. Likewise, exporters of traditional high-demand products such as coffee, cocoa, cashew nuts, nutmeg, mace, and coconut products such as Brazil, Columbia, Kenya, Jamaica, Cote d'Ivoire, Grenada, Indonesia, Malaysia, among others, have also faced similar regulatory and market-driven demands for the US, EU, Canadian, Japanese, and Australasian markets. Exporters of a range of traditional produce, canned foods, beverages, and sauces and spices as well as seafood exporters have also all had to deal with changes regulatory requirements in these markets, in addition to consumer-driven, significantly more demanding FSQS standards from the private importers and retailers, led by the major multiples and manufacturers.

Driven by the increased visibility and importance of foodborne illness outbreaks, the finding of filth (extraneous matter), as well as heavy metal, pesticides, and other residues in foods, misbranding, and a high incident of noncompliant or misleading labels, regulators have tightened scrutiny at ports of entry. These changing requirements for accessing metropolitan markets, as well as the insistence from the penultimate and final buyers in the distribution chain—the retail outlets and the consumers—on tougher, more effective traceability and food safety requirements to protect them have meant that traditional importers and their sources of supply have often been impacted, typically negatively. These importers would not have had the capabilities to keep updated with, and interpret the new market entry requirements for their suppliers, often resulting in significantly higher rejections at port of entry with the attendant losses occasioned by this. Conversely, the larger or newer importers and the multiples would have the resources to support their suppliers, ensuring continuity of supplies as changes in the requirements came into being.

Throughout this volume and the course of this chapter, we will examine the impact of the changing FSQS requirements in the United States, Canada, the European Union and other territories on the supply chain as well as their impact on exporting countries and individual exporters. The importance of controlling extraneous matter in foods, ensuring food safety and compliance with quality requirements and understanding and seeking to attain HACCP or GFSI-benchmarked certification will be explored. The approaches to addressing market access challenges and the nature of the resources required to effectively do this as part of the exporting endeavor is discussed. Foods from the Asia, the Middle East, Central and South America, the Caribbean and Africa are discussed. Examples of issues affecting Mexico (mangoes and papaya), Indonesia (nutmeg and mace), Jamaica (ackee, queen conch, jerk seasoning), and Asian, African, Middle Eastern, Lain American, and Caribbean sauces and spices and other traditional products (coconut and coconut products, selected seafood, formula safe foods, tropical juices and other beverages, vegetables) are also discussed. The chapter will set the stage for detailed discussions presented through other, more extensive case studies, as to how FSQS strategies can and have been deployed to mitigate, reduce, correct, or prevent existing and potential problems arising from the contamination of foods or noncompliance with FSQS requirements from governments or the target market.

Sources of contamination in foods and foodborne illnesses

Foodborne illnesses arise from a variety of causes, some of which are among the more innocuous, so much so as to appear benign, while others are more obvious and predictable. Whatever the source, steps need to be taken to secure the safety of the foods being grown, handled, processed, and delivered to tables around the world. Among the contributors to foodborne illnesses is foreign matter which come from a variety of sources in the production, manufacturing and even the use environment and can end up in food being sold to consumers. Foreign matter (also called extraneous matter and filth), in this context, is anything that ends up in food that ought not to be there, not being a natural and wholesome part of that food. Foreign (extraneous) matter in food is examined in more detail in the subsequent section, as also is its relationship to foodborne illnesses.

Foreign matter in foods

Food can be contaminated from a variety of sources depending on how it is produced, the nature of its environment and how it is handled and stored. Potential contaminants include pests, foreign matter, chemicals, and microorganisms. Pests include birds, insects (flies and cockroaches), and other animals, such as lizards and rodents. If not controlled, these pests can either contaminate food directly or with the microorganisms that they carry on their bodies. As indicated previously, foreign matter is anything that ends up in food that does not belong there, and comes from a variety of sources. It includes items such as hair, nails, glass fragments, stones, plastic and metal, all of which can cause physical injury or trauma. A list of the some of the more common foreign matter found in food is presented in Table 2.1. Examples of other types

Table 2.1 Some examples of foreign (extraneous) matter and their sources

Material	Potential injury	Sources
Glass	Cuts, bleeding; may require surgery to find or remove	Bottles, jars, light fixtures, utensils, gauge covers
Wood	Cuts, infection, choking; may require surgery to remove or find	Fields, pallets, boxes, buildings
Stones	Choking, broken teeth	Fields, buildings
Bullets/BB shot/ needles	Cuts, infection; may require surgery to remove	Animals shot in field, hypodermic needles used for infections
Jewelry	Cuts, infection; may require surgery to remove	Employees
Metal	Cuts, infection; may require surgery to remove	Machinery, fields, wire, employees
Insects and other filth	Illness, trauma, choking	Fields, plant postprocess entry
Insulation	Choking, long-term illness if asbestos	Building materials
Bone	Choking, trauma	Fields, improper processing
Plastic	Choking, cuts, infection; may require surgery to find or remove	Fields, plant packaging materials, pallets, employees
Personal effects	Choking, cuts, broken teeth; may require surgery to find or remove	Pens/pencils, buttons, careless employee practices

Figure 2.2 Contaminants found in foods.
Source: André Gordon, 2016.

of foreign matter that have been found in our[b] laboratory while examining foods from around the world are shown in Fig. 2.2. These range from various types and sizes of fibrous material Fig. 2.2A, larvae (Fig. 2.2B) and live insects (Fig. 2.2D) to small rodents and reptiles (Fig. 2.2C).

Food can be contaminated at any stage of production, handling, processing, distribution, selling, at retail or while being stored and subsequently used in the home. Among the many sources of contamination for products of agricultural origin, grown on farmlands are:

- animals
- soil
- air
- pesticides and fertilizers
- chemicals applied to crops
- soil amendments
- postharvest handling/transport
- Environment
- equipment/trucks
- packaging

Figure 2.3 Examples of the use of water in a packing house.
Source: André Gordon, 2006.

Another major source of contamination is the workers on the farm or in the packing house and water. The water may be used for:

- irrigation (application to the crops in the field),
- during immediate postharvest handling (including removing soil from the harvested product) and
- during packing (for transfer of fruits/vegetables around the facility, washing, etc.; Fig. 2.3).

Consideration should also be given to other sources of extraneous matter that may contaminate product, including but not limited to the packing house environment, secondary and tertiary packaging materials, chemicals and packing house equipment. For products manufactured in a food processing plant, many of the same concerns and sources of contaminants exist but there are also several others, such as chemicals associated with cleaning and sanitation, chemicals associated with lubrication of equipment and, those involved with the treatment of pests. There are also specific concerns about foreign matter and microorganisms that could get into the product because of the mishandling at the plant level or which may have come in with one or other of the ingredients. The handling of these scenarios is explored in this book.

Because of the multiplicity of options for contaminants to get into food, it is important that effective control is exercised all along the production chain to exclude them. This means that in handling products of agricultural origin, including items such as peppers, efforts must be made to exclude or find and remove all insects (Fig. 2.2D) and other contaminants that may be naturally present, prior to packing or processing. It also means that pest control, a very important prerequisite program (PRP), has to be very effective or the outcome can be disastrous. This was so in several instances that were handled in our laboratory,[c] one such being the finding of a partially eaten lizard in a ready-to-eat (RTE) meal which was the focus of a consumer complaint (Fig. 2.2C). This became the subject of a lawsuit in which the claimant was successful, her daughter having to be hospitalized after reacting to finding that she had partially eaten the lizard. In this particular case, the RTE meal was imported and, en route to establishing plausible liability, we[d] had to prove that the reptile originated in the region from which the item was imported, and not in the country of consumption.

In another case, a partially consumed rodent was found in a meal from a quick serve restaurant (QSR). Again, the outcome was not favorable for the QSR. Incidents such as these, if widely publicized can cause irreparable damage to a product or brand. It is therefore imperative that firms must ensure that they have adequate PRPs in place to eliminate the likelihood of foreign matter ending up in their product. An example of the structure of an effective PRP is given later on in the chapter and the implementation of PRPs is discussed in several of the case studies in this book.

Selected cases of food poisoning of note for developing countries

The impact of foodborne illnesses associated with a product from any company in any region or country can be even more problematic. The situation has been exacerbated by the emergence of more recent and aggressive pathogens and the ubiquity of telecommunications technology that allows for the instant circulation of information globally. This means that whenever there is an outbreak, even before the actual source is confirmed, the information on the product, the company, the region or country in which it has occurred and the supposed cause will likely be not only on the major news channels, but on a firm's potential consumers personal communication devices (tablets, cell phones) within hours, if not minutes. Major pathogens such as *Escherichia coli* O157H7, *E. coli* O104H4, *Clostridium perfringens*, *Campylobacter jejuni* and *Campylobacter coli*, *Listeria monocytogenes* and *Salmonella* [Centers for Disease Control and Prevention (CDC), 2016b], as well as noroviruses, Hepatitis A and rotaviruses [European Food Information Council (EUFIC), 2014] are creating significant challenges for the global food supply chain. This is despite significant efforts by firms and countries to implement programs to eliminate, minimize, or control their presence in the food supply. In the case of firms from developing countries, many of whom may have less resources available to them than their developed country counterparts, the results will likely be disastrous, as has often been the case. These firms must therefore ensure that effective systems are in place to mitigate against this threat to their business. And this threat has arisen in some unexpected circumstances and with unexpected products, with disastrous results.

Among the more notorious cases of illness caused by the innocuous use and consumption of a daily staple, water, was the outbreak of *E. coli* and *Campylobacter* poisoning that occurred in Walkerton, Ontario, Canada in 2000. Drinking water from their municipal supply that was contaminated with fecal matter from nearby farms resulted in the death of seven people and over 2500 illnesses (Clark et al., 2003). In this outbreak which was caused by *E. coli* O157:H7 and *Campylobacter* spp., the food industry experienced its first globally publicized case of *E. coli* O157H7, which made the organism a focus because of the impact on survivors. *E. coli* O157:H7 attacked the organs of the many of the victims of the waterborne outbreak, particularly their kidneys, resulting in a disease called hemolytic uremic syndrome (HUS) which is characterized by hemolytic anemia caused by the destruction of red blood cells, uremia[e] and thrombocytopenia.[f] Many of the victims who survived remained seriously ill and in need of dialysis.[g] The scale, scope, and impact of this outbreak caused renewed focus from the industry, the media and the public not just on this particular strain of

E. coli (O157:H7), first isolated at an outbreak associated with beef patties (Riley et al., 1983), but also on the issue of the impact of foodborne illnesses in general which has continued up to today. Subsequent highly publicized outbreaks have kept the issue on top of the mind and has changed the nature of the food industry such that food safety will remain a priority for some time to come.

Selected foodborne outbreaks linked to developing countries

The focus on food safety has been heightened as there has been an evolution in the types of foods from which outbreaks occur and also, critically, the organisms now being found on, and associated with these foods. As the demand and market for food from, and of the type typically grown in, developing countries has rapidly expanded, so too has the foods from which foodborne illness can occur been evolving, increasing the focus on these and other foods as potential causes of outbreaks. When the US Food and Drug Administration had to move to introduce a mandatory requirement for HACCP systems in juices through 21 CFR Part 120 (Food and Drug Administration, 2001) because, among other things, of the *E. coli* O157H7 food poisoning incident with New England–produced apple cider (Besser et al., 1993), it marked, to some extent, the beginning of the end of the age of innocence regarding foods previously thought to be immune to the risk of contamination and survival of such well-known pathogens. So too has this and other incidents enhanced our understanding of what causes the contamination of these products to occur, the survival of the pathogens in them and how to deal with them (Leyer et al., 1995). For developing countries, while many of their tropical fruits, vegetables in particular, and other foods are now being highly sought after, the veneer of automatic food safety associated with them has been shattered by some important cases of note over the last decade or so which bear examination. This will allow the lessons learned to advise future action, hence the following discussion.

Escherichia coli O104H4 on fenugreek sprouts

Originally thought to have been from fenugreek sprouts seeds imported from Egypt, a major outbreak of food poisoning due to *E. coli* occurred in Germany between May and June 2011. The German authorities identified Egypt as being the original source of the seeds for the sprouts which were found to have been produced by an organic farm in Bienenbüttel in Lower Saxony, Germany. The ensuing outbreak resulted in more than 4000 people in 16 countries becoming ill, requiring international surveillance to determine the scope of the outbreak [Centers for Disease Control and Prevention (CDC), 2013]. There were 15 cases of HUS in France resulting from the outbreak (Gault et al., 2011), 6 cases in 5 states in the United States of America, 35 cases in Sweden, 15 in Denmark and cases in Canada, Norway, Luxembourg, Spain, and the United Kingdom, among other countries. As a result of the outbreak, 53 persons died. Over 800 person were afflicted with HUS, at least 33 of which remain on dialysis for the rest of their lives. Hundreds of millions of dollars were lost by the produce sector, particularly cucumbers, lettuce, and tomatoes, which were originally identified as possible causes, resulting in many countries imposing restriction on imports and the industry, particularly in Spain, having to dump over US$200 million in produce. Egypt, the source of the sprout seeds, also found itself facing restrictions on its exports.

The outbreak was originally thought to be caused by enterohemorrhagic *E. coli* (EHEC) because it was characterized by a large number of cases with complications, including HUS. It was eventually shown to be due to a novel strain of *E. coli*, *E coli* O104H4 (Qin et al., 2011), an enteroaggregative *E. coli* (EAEC) that had acquired the ability to produce Shiga toxin, apparently by lateral transfer (Casey et al., 2011). It was one of the most aggressive and costly outbreaks to have occurred and led to a significant tightening in food safety requirements across all affected territories. Critically, while Egypt maintained that it was not the source of the problem and that it had no cases of patients infected with *E. coli*, its systems of traceability and surveillance at the time, were not sufficient to make an irrefutable argument against the evidence coming out of Europe. This highlights the importance of developing countries having traceability systems mandated for exporters because of the impact of food safety challenges on the overall exports from a country. It also highlights the importance of these systems for the exporters themselves to mitigate the impact on their businesses. Critically, in evaluating this case, Rubino et al. (2011) noted

While Germany is not a developing nation, the handling of the outbreak can be used as a model to build a strategic plan for developing nations. First a well-coordinated network that shares information on outbreaks in developing countries will be invaluable. Networks help bring expertise and technology to resource-limited areas in a timely manner.

It is therefore essential that all developing countries and producers and exporters from these countries work together to ensure that their postharvest handling systems and systems of traceability, as well as their collaboration with other countries and partners give them the ability to effectively trace all products destined for third country markets.

Outbreaks with other fresh fruits and vegetables

Over the years, there have been cases of foodborne illness outbreaks associated with tropical fruits, caused mainly by *Salmonella* and norovirus as the primary microbial pathogens (Rothschild, 2011). This has included 26 cases of infection with *Salmonella* Litchfield in 2006 and 11 cases of *Salmonella* Saintpaul in 2009, both in Australia and both associated with papaya (called pawpaw in some parts of the world). In 2010, there were at least nine cases of *Salmonella typhi*–like infections in California and Nevada in the USA, suspected to be caused by another contaminated tropical fruit, mamey (*Pouteria sapota*) fruit pulp (Rothschild, 2011). Subsequently, there have been major outbreaks associated with fruits, both tropical and otherwise, that now occupy pride of place in most supermarkets (Fig. 2.4). These outbreaks have been caused by fruits, typically from the same source but widely distributed across states and countries. This has dramatically changed the way food safety of fruits is thought of and now demands a greater understanding of the issue if trade in these products is to continue to grow

Figure 2.4 Fruits being displayed for sale in Albert Heijn.
Source: André Gordon, 2016.

and if food safety practitioners are to develop and implement effective programs to ensure their safety. Three such outbreaks are examined in subsequent sections.

Jensen Farms *Listeria* outbreak in cantaloupes

The Jensen Farms *Listeria* outbreak in 2011 resulted in 147 illnesses in 28 states and killed 33 in the United States in October 2011 [Centers for Disease Control and Prevention (CDC), 2012]. This was the deadliest foodborne illness outbreak in the United States since 1924. The outbreak was caused by cantaloupes contaminated with five subtypes of *L. monocytogenes* that were grown and packaged by Jensen Farms in Colorado and shipped to many large retail entities across the United States which unknowingly sold them to customers. The firm, which had shortly before the incident been audited and recertified with a "superior" rating to a food safety system (Food Safety News, 2014; Marler, 2014), had been in the business for many years without having previously had a problem. Pooled water on the floor of the packing house, poor equipment design and work surfaces contaminated with *L. monocytogenes* were among the likely causes of the contamination (Food and Drug Administration, 2014), with inadequate sanitation and handling practices resulting in the contamination of the cantaloupes. This outbreak brought attention to the fact that fruits, which were previously thought by the general public to be safe and therefore excluded from the foods about which there needed to be concern for food safety, could no longer be considered "safe." It also brought into focus the relevance of food safety certification and the importance of thoroughness by auditors responsible for food safety certification (Food Safety News, 2014) and the issue of the handling practices in packing houses. In addition to the death and ongoing illness for some victims, the incident remains the subject of lawsuits amounting to hundreds of millions of US dollars, brought Jensen Farms into bankruptcy and earned sentences for its principals. It continues to serve as a lesson for how RTE fruits and vegetables should not be handled and the consequences of mishandling.

Food poisoning from papayas

In July 2011, a major outbreak of foodborne illness due to *Salmonella* Agona was reported in the United States. The source of the illness was fresh, whole papaya

Figure 2.5 Mangoes on a tree (A) and papaya (pawpaw) (B).
Source: André Gordon, 2016.

(as in Fig. 2.5B), imported by Agromond Produce of McAllen Texas, USA from Mexico (Marler, 2011; Rothschild, 2011). Initially, reports were confined to the United States but the Centers for Disease Control and Prevention (CDC) later confirmed cases in Canada due to the importation of the contaminated produce into that country by Agromond Produce [Centers for Disease Control and Prevention (CDC), 2011]. In the final assessment, the FDA and the CDC found that the outbreak resulted in 106 illnesses across 25 states in the United States and Canada, with victims who suffered from symptoms including diarrhea, fever, and abdominal cramps ranging in age from 1- to 91-years old. The papayas were distributed under four brand names—Blondie, Yaya, Mañanita, and Tastylicious and went to major distributors across the United States and Canada, including Walmart. This was second such major outbreak from *Salmonella* in 2 years associated with fresh fruit, the previous one in 2010 having led to 119 cases of illness across 14 states. The specific source of the illness was never determined [Centers for Disease Control and Prevention (CDC), 2011].

Food poisoning from mangoes

In August 2012 a major recall of mangoes (Fig. 2.5A) took place across the United States because of 127 illnesses across 15 states and 22 cases in Canada caused by contamination of mangoes with *Salmonella* Braenderup. Although no deaths were recorded, 33 people were hospitalized. The Daniella-branded mangoes were imported into the United States from Agricola Daniella of Sinaloa Mexico and distributed widely throughout the affected states through retail outlets including Metro Market, Pick'n Save, Stop & Shop, and Costco. The mangoes were imported by Charlie's Produce in Washington State and Splendid Products from California [Bottemiller, 2012; Centers for Disease Control and Prevention (CDC), 2012]. This outbreak was the third major outbreak of food poisoning from mangoes, the first being caused by *Salmonella* Newport and *E. coli* in 1999 from Brazilian mangoes, and the previous one being from *Salmonella* Saintpaul contamination of mangoes, some of which were from Peru, in 2001. In both cases, nonchlorinated water used for washing the fruit was implicated (Marler, 2012).

Foodborne illness outbreak from the parasite Cyclospora

Before the mid-1990s, very few persons in the developed world had heard about *Cyclospora cayetanensis* as this parasite was largely associated with waterborne gastroenteritis among poor children living under unacceptable sanitary conditions or adults who may have drank untreated water of questionable quality in developing countries where the parasite was endemic. This changed dramatically in the late 1990s when outbreaks of cyclosporiasis, the disease caused by the parasite, and eventually traced to Guatemalan raspberries occurred in 1996 and 1997 in the United States [Bern et al., 1999; Centers for Disease Control and Prevention (CDC), 1997] and in 1998 in Canada, with an outbreak from fresh Guatemalan blackberries also in Canada in 1999 [Canadian Food Inspection Agency (CFIA), 2016]. By the time the source was identified and appropriate interventions taken, there were more than 1465 cases across 20 states in the United States and two provinces in Canada, with just under 50% (actually 49.5%) of the cases associated with the consumption of tainted raspberries at events across the continent [Herwaldt and Ackers, 1997; Centers for Disease Control and Prevention (CDC), 1997]. The events included multiple persons having meals together at the same restaurants, reception, and banquets. The outbreaks resulted in 22 hospitalizations, but no deaths (Herwaldt and Ackers, 1997). The scale and impact of this outbreak, which was the largest ever of its kind, established *Cyclospora* as foodborne illness vector of interest and dramatically changed the way the global fresh fruit, vegetable, and produce business operated.

Cyclospora is a coccidian parasite that can be found in untreated waters or water of poor sanitary quality, in which it would typically be present as oocysts. The oocysts are the resistant stage of coccidian parasites that allow them to survive in unfavorable conditions, releasing the infectious sporozoites when conditions become favorable. Cyclosporiasis, the disease caused by *Cyclospora*, has an average period of incubation of 1 week, is associated with invasion of the small intestine by the sporozoites[h] of the parasite and is manifested by gastroenteritis treatable with trimethoprim–sulfamethoxazole. If not treated promptly, symptoms can be protracted, remitting, and relapsing. Like the oocysts of other coccidian parasites, *Cyclospora* oocysts, can be destroyed by pasteurization or commercial freezing processes, although the minimum time and temperature conditions required to inactivate the *Cyclospora* oocysts may vary (Ortega and Sanchez, 2010).

C. cayetanensis, the causative agent in the outbreak, is a pathogen commonly associated with pediatric gastroenteritis in Guatemala, especially from May through August, the rainy season, a pattern of seasonality that is similar to that observed also in Kathmandu (Bern et al., 1999). First reported in Guatemala in 1994 (Herwaldt and Ackers, 1997), cyclosporiasis occurs when untreated waters contaminated with feces from an infected person is consumed by unsuspecting susceptible individuals. Although the actual source of the *Cyclospora* was not definitively established, it was thought most likely to have come from the use of untreated water to mix fungicides sprayed directly onto the fruits (Bern et al., 1999). This is because direct fecal contamination by handling is an unlikely source due to the incubation period required for the organism to become infective if not already in the appropriate form (Ortega et al., 1993).

While the outbreak was initially thought to be due to strawberries from California, it was eventually linked to Guatemalan raspberries close to the end of the growing season in 1996. It was not therefore possible for the FDA to be definitive about the source and so exports continued in 1997, when again, an outbreak ensued (Calvin et al., 2003). The scale of the outbreaks resulted in the FDA requesting that the government of Guatemala (GOG) and the Guatemalan Berries Commission (GBC) voluntarily suspend shipments in May 1997, with a view to the FDA, the Centers for Disease Prevention and Control (CDC) and GBC and the government of Guatemala working together to solve the problem. However, by December 1997, not convinced that the problem had been effectively addressed, the FDA imposed an import alert on all raspberries coming from Guatemala. To regain access to the US market, the GBC and GOG had to develop and put effective measures in place to prevent a recurrence. Working with the FDA, the CDC, the CFIA, Health Canada, and the Food Marketing Institute (FMI) in the United States which represents US retail buyers, the GOG and GBC developed and implemented the Marketing Plan of Excellence (MPE) by 1999. A mandatory joint program of the GBC and GOG, the MPE[i] required Guatemalan growers of raspberries and blackberries to implement stringent food safety programs, full traceability, and undergo FDA inspections (Calvin et al., 2003). In 1999, based on its comfort with the program, the FDA lifted the import alert on Guatemalan berries and allowed the trade to resume. In 2000, there were two further outbreaks of cyclosporiasis associated with Guatemalan raspberries in 2000. One delinquent farm was removed from the MPE and there has been no further outbreak since then.

To put this outbreak in context, one has to revisit the circumstances surrounding it and the impact it had on Guatemala. The local raspberry industry had grown rapidly by capitalizing on low supplies of raspberries in the United States during the spring and fall. In 1996 when the first outbreak occurred, the Guatemalan industry was experiencing unprecedented growth with about 85 growers involved and year-on-year growth of over 110% (Calvin et al., 2003). When the problem arose, the GOG and the GBC reacted slowly, to characterize the risk, develop a plan, and identify potential problem farms. Further, there was no traceback to implicated farms, nor was there an enforcement mechanism. By 2001, Guatemalan exports of raspberries was only 16% of the 1996 levels and by 2002, only three farms remained as active exporters (Calvin et al., 2003). Today, Mexico and Chile dominate US raspberry imports, a market that once created better livelihoods for Guatemalan farmers and which Guatemala once thought it would own (Flynn, 2013).

In an excellent summary of the impact of the Guatemalan raspberry *Cyclospora* outbreaks in 1996 and 1997, Calvin et al. (2003) noted:

> The Guatemalan problem with Cyclospora was a critical event in the produce industry. Producers everywhere noted the devastating impact a food safety problem could have on an entire industry and learned important lessons:
>
> (1) delay in addressing such a problem may adversely affect an industry's exports and reputation;
> (2) the FDA may make decisions on trade restrictions based on epidemiological evidence alone without physical evidence;

(3) improved traceback allows trade restrictions to be targeted at individuals with contamination problems and not at the entire industry; and

(4) strong grower organizations can improve an industry's ability to deal with food safety outbreaks.

When the California strawberry industry was initially and incorrectly implicated in the 1996 outbreak, Guatemalan growers saw the California Strawberry Commission respond quickly and strongly to the negative publicity. The GBC learned from that experience and has significantly improved its ability to deal with such a situation, should one occur in the future.

Outbreaks such as these affecting products from Egypt, Brazil, Peru, Mexico, and Guatemala, as well as other outbreaks that have originated in developing countries are a cause for concern for the affected region and exporting countries. They could stigmatize fresh fruit and other food exports and hurt what has become a growing trade for these countries, as occurred in Guatemala. Consequently, food safety is now a major consideration for exporters of food and agricultural produce from developing countries. This is because these outbreaks have completely changed the game for imported fresh fruits and vegetables and have resulted in the major buyers across the world demanding evidence of the safety of the products they import through a variety of means, including the certifications systems discussed later. Fruits and vegetables such as those mentioned previously, as well as other that have such significant potential for growth, including ambarella, soursop, peppers, and callaloo, which were discussed in Volume I (Gordon, 2015a), must now ensure that they fully comply with industry best practices (typically, including GAPs). They must also comply with the relevant import regulations, including the EU directives, the SFCA, the preventive controls, foreign supplier verification program (FSVP), and produce rules (Food and Drug Administration, 2015a,b,c) which are discussed in more detail throughout this volume.

Meeting market requirements

Major buyers such as Loblaws, Costco, Whole Foods (North America), Tesco, Sainsbury, Asda, Carrefour, Albert Heijn (European Union), and others are showing increasing interest in carrying authentic products from various parts of the developing world such as Africa, Latin America, the Caribbean, and Asia. This means that these products, if properly presented (Fig. 2.6), have a good chance of being commercially successful in Europe, North America, and Australasia as well as some newly emerging markets, including China, Brazil, and parts of Central and Eastern Europe. To capitalize on these opportunities, producers and exporters have to ensure that they are able to meet the requirements to access the countries where the opportunities have been identified. This includes ensuring that their technical and production teams are properly trained in the requirements for exporting to the markets targeted, having appropriate GMPs and PRPs in place and understanding the regulations for the markets into which they are exporting. These include requirements such as the EU Regulations (EC) No. 178/2002 and No. 852/2004 for exports to Europe and the Safe Food for Canadians

Figure 2.6 Traditional oriental foods packaged for developed country markets.
Source: André Gordon, 2016.

Act (SFCA) for exports to Canada. They also have to document the shelf life of products going into these third country markets and get the production facility and process fully EU, FDA, and SFCA compliant to ensure continued market access.

If they are already exporting to the US market, it means understanding the Preventive Controls (Hazard Analysis and Risk-based Preventive Controls) rules, the Foreign Supplier Verification Program (FSVP) and the Produce Act (Food and Drug Administration, 2015a,b,c) and other FDA regulations which are relevant to their products (Saltsman and Gordon, 2015). Producers and exporters also have to ensure that they are able to verify the efficacy of their GMPs, PRP, production processes, and process controls, that their labeling, including the declarations on the labels, are fully compliant and that they have and are able to present scientifically valid data supporting their traditional processes. Exporting firms therefore now need to ensure that they understand and have technical support and capabilities to interpret and fully comply with regulatory requirements if they are to retain their existing markets or to break into new markets around the world.

In many of these more developed markets, an important aspect of successful market penetration and growth is the attainment of certification, not just HACCP certification but Global Food Safety Initiative (GFSI) benchmarked certification, as discussed previously. The GFSI is a comprehensive grouping of food industry interests that joined together in 2007 to standardize FSQS certification globally, creating an industry driven collaborative platform to advance food safety (Global Food Safety Initiative, 2016a). Among the major food business that subscribe to and endorse the

Figure 2.7 Current members of the Global Food Safety Initiative Board, 2016.
Source: http://www.mygfsi.com/news-resources/news/471-amazon-tesco-auchan-and-dole-join-the-gfsi-board.html.

GFSI are Groupe Carrefour, Shoprite, Campbell's, Migros, Daymon Worldwide, Hormel, ASDA, US FoodService, COOP, Hormel, Aeon, the Coca Cola Company, the Delhaize Group, Kraft, ConAgra, ICA, Metro Group, Tesco, Loblaws, Publix, Danone, Sodexo, and Tyson. Current members of the GFSI board include the major multiples, such as Walmart and Groupe Carrefour as well as transnationals such as Dole, Cargill, and McDonald's (Fig. 2.7). Understanding the business culture of each production enterprise and choosing the right systems for the business is critical to getting and maintaining certification to a GFSI-recognized FSQS standard (such as SQF, BRC, FSSC 22000, GlobalG.A.P.). The GFSI, the differences between the major benchmarked schemes and guidelines for successful implementation will be discussed in Volume III of this series but further information can be found at their website.[j] For now it is important to note that major buyers have already given deadlines for exiting suppliers to attain and maintain GFSI certification and often require these and other certifications prior to engaging a new customer or product. This means that all potential supplier firms need to upgrade their FSQS to a GFSI-benchmarked, HACCP-based FSQS that is one of those accepted by the GFSI, and be able to demonstrate, by way of certification, effective implementation and maintenance of these systems.

To deal with all of the current and future considerations, emerging markets and developing country food industry stakeholders have to ensure that they not only understand the imperatives driving these changes, but put themselves in a position to effectively handle them. This book will explore a wide variety of situations that have arisen relative to trade with developed country markets for products coming from developing countries and the ways that they can be, or have been dealt with effectively through the application of food science and technology. A significant focus will be on the application of food science to provide effective solutions for challenges that have arisen and also to the implementation of world-class, certifiable, and certified FSQS. This will be explored largely by way of case studies and examining some important firm-level and national challenges that have arisen and how these were dealt with. The cases and situations examined are relevant in all countries and to all situations. Three such short case studies are discussed in subsequent sections.

Addressing trade and market access issues

The first volume of this series placed significant focus on the case study of Jamaican ackees and the approaches taken to gain reentry into the US market (Gordon et al., 2015) as well as to prevent its prohibition from the Canadian and UK markets (Gordon, 2015a). In the latter chapter, in examining the issue of dealing with trade challenges, Table 7.2 presented a range of issues that had been encountered by the author's firm, Technological Solutions Limited (TSL) and successfully dealt with. Some of these, including issues with canned callaloo (*Amaranthus*), Irish Moss (a *Gracilaria* beverage) and canned pasteurized process cheese will be examined in more detail later in this volume. Some, such as regaining access to the EU market for Queen Conch (*Strombus giga*) which will be discussed in more detail later, like with ackees (summarized later), involved industry-wide systems implementation and legislative and regulatory reform. The others, such as those summarized in Table 2.2 (extracted from Table 7.2; Gordon, 2015a) are also explored in subsequent sections.

Jamaica's canned ackee

As with the circumstances surrounding the Guatemalan raspberry industry mentioned earlier, Jamaica had an import alert imposed on its exports of canned ackees (*Blighia sapida*) and then ackees in any form in 1973. However, the Import Alert on Canned Ackee exports from Jamaica to the United States lasted from 1973 to 2000, as was discussed in detail in Gordon, (2015a), and was only lifted when a comprehensive program similar in some ways to what was deployed in Guatemala was applied in Jamaica in 1999/2000. This private sector Jamaica Ackee Task Force[k]-led program[l] was far more comprehensive and ambitious in scope because of the complexities involved with a product, ackee, on which the import alert was implemented because it contained a natural toxin that required stringent control. Further, unlike the Guatemalan situation, the toxin caused an illness, toxic hypoglycemic syndrome (THS), characterized by hypoglycemia, which could be fatal. Also, there was no organized growing of the fruit and although there was much anecdotal evidence that it could be handled safely, the research to date did not provide strong enough scientific support to give the FDA comfort. To regain access after a 27 year absence from the market because of the import alert, the industry had to be transformed for a quality systems/standard of identity-based regulatory model and practice to a food safety–based production system based on HACCP principles. This was augmented by a national program of changing handling practices across the industry, retraining all

Table 2.2 **Selected Caribbean exports to developed country markets with market access issues due to technical and scientific issues and their mitigation**

Product	Exporting and importing country	Nature of the issue/challenge	Approach to resolution
Crackers (biscuits)	Trinidad and Tobago/the United States	Noncompliant labeling/Proof of Certification of FD&C colors	Amended label to become compliant. Obtained FDA certification of all batches of FD&C colors used in the food. Produced as evidence; kept on file
Red Pepper Sauce	Jamaica/the United Kingdom	Disallowed color (Sudan Red) found in product	Source of banned dye identified by analyses as being admixed Central American pepper mash; removed
Mango kuchela (*M. indica* sauce)	Trinidad and Tobago/the United States	Product not filed with CFSAN[a]; formula safety of product not proven	Shelf stability and safety of product established through studies and thermal process validation and filed with CFSAN's LACF division

[a]The United States Food and Drug Administration's Center for Food Safety and Applied Nutrition (CFSAN).
Source: Adapted from Gordon, A., 2015b. Dealing with trade challenges: science-based solutions to market-access interruption. In: Gordon, A. (Ed.), Food Safety and Quality Systems in Developing Countries: Volume One: Export Challenges and Implementation Strategies, Academic Press, pp. 115–128; Table 7.1, where these challenges were originally handled by Technological Solutions Limited (TSL), the author's company.

value chain participants in GMP and food safety, retraining the inspectorate arm of the government to operate according to food safety (HACCP)-based inspection principles and doing the required scientific research to provide the backing for the new approach to regulating the industry (Gordon et al., 2015). The result of this successful program, unlike in the Guatemalan situation, was that ackee exports moved from US $4.4 million in 1999 prior to the lifting of the import alert to US $15.4 million in 2015, based on the same volume of exports. A comparison of the two approaches, circumstances and outcomes might be instructive.

Labeling and the use of certified colors

As reported by the Economic Commission on Latin America and the Caribbean (ECLAC) in its review of trade in 2011 (ECLAC, 2011), labeling continues to be one of the major causes of rejection of exports from the region to the United States. As for domestically manufactured food products, all exports into the United States are required to comply with all applicable labeling regulations, including the following:

21 CFR 101—Food Labeling
21 CFR 73, 74 and
21 CFR 82—Labeling of Certified Colors
The Nutrition Labeling and Education Act (NLEA)
The Food Allergen Labeling and Consumer Protection Act (FALCPA)

All of these, found in the Federal Register[m] or from the FDA's website[n] outline the requirements for foods to be traded in the United States to be compliant with all labeling regulations. It is the contravention of these regulations that have resulted in the 16% rejection of products at the US ports of entry for labeling breaches (ECLAC, 2011). Exporters are required to comply with the general food labeling regulations 21 CFR 101, which include a section 21 CFR 101.9 which deals with Nutrition Labeling and which has been updated over the last several months. Compliance is also required with the Nutrition Labeling and Education Act (NLEA) which sets out the requirements for the display of nutrition information on all foods labels and which also is being updated. Allergen labeling is controlled under the FALCPA. While many people are aware of these, very few were aware of the regulation that requires the labeling of certified colors, 21 CFR parts 74 and 82.

An example of the kind of problem that can arise for developing country exporters not cognizant of the requirements is the challenge that arose for an exporter of baked goods to the United States, specifically biscuits and snacks from Trinidad and Tobago in the Caribbean. This firm which has a moderate, fairly modern plant supported by a good technical team had been producing and exporting its products to the United States for years without a problem. In the case indicated in Table 2.2, a shipment of crackers was held at the New York port of entry in 2013 and examined for compliance with US regulations. It was found that the label did not have the appropriate nutrition labeling format, queries were raised about the accuracy of the allergen labeling declaration and, perhaps most surprisingly and (difficult to address), *the firm was asked to provide the certification numbers for the batches of certified colors declared on its label* and used in the manufacture of lots of product in question. The firm approached TSL which assisted them with addressing the challenge.

The nutrition labeling issue was addressed by comparison of the declaration with the composition of the product and providing this to the FDA. The allergen declaration was changed to capture other possible allergens that, while not in the product, were also handled in the plant and therefore could be argued to be a potential risk for cross contamination. As such, the declaration was adjudged to be OK and so indicated to the FDA, who accepted that position. The situation regarding the colors was different. This was a new request and prompted detailed examination of the regulations governing certified colors. 21 CFR 73 provides information on "Color additives exempt from batch certification"[o] while 21 CFR 74 provides information on "Color additives subject to batch certification,"[o] both of which outline those colors which are exempt from and which require batch certification by the FDA, respectively. On the other hand, 21 CFR 82 provides the "Listing of Certified Provisionally Listed

Colors and Specifications" which details which colors are listed and how they are to be designated.[p] Collectively, these indicated that the manufacturer needs to get from their supplier of FDA certified FD&C colors the relevant certificate numbers that verify to the FDA, if asked, that the colors being used were from a batch that was actually certified by them.

In this instance, this is what was done, with some difficulty as the supplies of the colors were not aware that this was required and, themselves not being the makers of the colors had to source this from their suppliers. The manufacturer now routinely requests the certificate numbers from the suppliers of certified colors and does not buy from anyone who cannot supply this information. This represented one of those circumstances when the 2–4% of containers entering US domestic trade that get inspected raised an issue that had not been dealt with before and which provided an opportunity for learning and will prevent the firm from having such challenges in the future.

Undeclared ingredient in products exported to the United Kingdom

Another situation that arose with one of TSL's clients and expanded to affect a whole industry segment across the Caribbean region is captured in Table 2.2 in the case of Red Pepper Sauce. This situation could also easily arise for many developing country exporters as the circumstances, to a large extent, were beyond the control of the manufacturer. In the case of Gray's Pepper Products,[q] a well-known pepper sauce producer and exporter with over 40 years in the business, the firm made and exported pepper sauce as it usually did to the United Kingdom in 2005. Unknown to the firm, a storm was brewing in the European Union and the United Kingdom in particular over the widespread use of a range of synthetic Sudan dyes which produced degradation products considered to be carcinogenic and teratogenic.[r] This was not an issue in the industry in the Caribbean region because the dyes were not traditionally used in Caribbean foods and so no one knew about them.

The firm's product was held at the port of entry, an unusual occurrence for that market at the time, and tested for the presence of Sudan Red, a dye that was popularly used in tomato, chili- and pepper-based products at the time to enhance their red color. Gray's was surprised at the detention for testing, but understood that this happened occasionally at ports of entry. They were not concerned because they had a long history of producing only high quality, all natural, fermentation-based pepper products. When they were informed that the shipment was being rejected because of the presence of the a dye, Sudan Red, they were not only shocked, but alarmed as not only had they never heard of the dye, they could not believe that it could be found in their product, not having added anything unusual to their base ingredient, pepper mash (crushed peppers). The finding of the dye in their product was therefore cause for consternation and they naturally challenged the finding by the UK authorities through their importer. Once it was unequivocally confirmed that Sudan Red was present, the firm had to deal not only with the issue of loss of the shipment but having to put a hold on all future shipments until it could figure out the source of the contamination of its product.

By this time, all shipments of pepper sauces and crushed pepper mash from the Caribbean into the United Kingdom and European Union were routinely being held

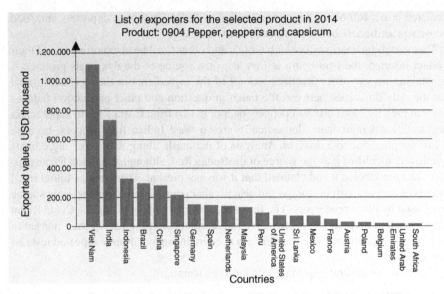

Figure 2.8 Selected developing country peppers and pepper products exporters, 2014.
Source: ITC Trademap (from the COMSTAT database).

and tested, along with similar shipments from around the world. Shipments with any of the Sudan or similar dyes were being rejected at port of entry, creating havoc in the billion dollar industry (Fig. 2.8). What started as a problem for one firm had now become a cessation of trade for all pepper sauce exporters from the region and many developing countries across the world. What was the cause of this "sudden" prohibition and where did the dye in Gray's and other Caribbean products come from if not added by the manufacturers?

The problem started in 2003 when the EU Rapid Alert System for Food and Feed (RASFF) issued a series of notifications about the presence of Sudan dyes in chili and other products (Hayenga, 2011). This led to immediate action, including product recalls and withdrawals and ultimately destruction of the affected products. In 2004 other related dyes were found in products in Germany and in February 2005, the United Kingdom issued a series of recalls for a wide range of foods with the illegal colors, these having been ruled unsafe by the European Union in 2004 by 2004/92/EC. The Sudan dyes (including Sudan Red B, Sudan Red G, Sudan Red 7B, Sudan Orange G and Sudan I to IV), along with other dyes like 4-(dimethylamino) azobenzene (Butter Yellow) are synthetically produced azo dyes whose degradation products had been regarded as unsafe for use in foods for some time. By mid-2005, the situation had escalated across the European Union because of the extent of the problem. This precipitated a review by The European Food Safety Authority (EUFSA) in August 2005 (European Union Food Safety Authority (EUFSA), 2005) of the toxicology of a group of dyes present in foods imported into the European Union, including the Sudan dyes, which it published and shared with national EU regulatory authorities. Consequently, by the time the Caribbean products were being held at UK ports, the situation had escalated into a full blown food safety crisis, with the Caribbean exporters and food industry practitioners being completely unaware.

This was the scenario into which Gray's and other Caribbean exporters found their product inserted, the firm being at loss as to the source of the dye in its product. A painstaking review of its practices and all of the ingredients in their product found that the only difference between the batch in question and other production batches was that they had used crushed pepper (pepper mash) from Costa Rica to supplement their traditional ingredient, domestically grown West Indian Red peppers, because of low supplies due to a drought. Analysis of the mash, along with other ingredients definitively identified it as the source of the Sudan Red, although the Costa Rican suppliers had not labeled it and claimed that it was not present. The dye was found in all batches of pepper mash in storage and also in other pepper mash from the same source being used by other processors. The firm subsequently stopped using the Costa Rican pepper mash in its product and was able to recommence shipments to their market in the United Kingdom, without the problem ever arising again, despite repeated tests on their product.

This case has several important lessons to be learned:

1. It is incumbent on, and critically important for all processors to have detailed knowledge about all of the ingredients being used in their products.
2. They need to insist that all suppliers provide full disclosure and detailed labeling information on ingredients being supplied. If they cannot do this, it is better to find other suppliers who can.
3. Processors, exporters, and the food industry practitioners associated with them should ensure that all raw materials are bought according to detailed raw material specifications, for which they provide certificates of analysis (COAs) with each batch or Letters of Guarantee (LOGs). This will provide both control and protection from situations such as occurred with the pepper sauce.
4. It is important that processors, exporters, and food industry practitioners keep themselves abreast of important technical issues affecting the food industry, in general, and their specific industry subsector in particular. Critically important as well is for them to keep updated with what is happening in the markets in which they do business. Had that been the case with the pepper sauce incident, the problem and subsequent losses might have been minimized.

Failure to learn from or be guided by the lessons aforementioned can be quite costly and may, depending on the scale of the problem and nature of the business, put the firm's business at risk.

Thermal process filing of a traditional naturally acidic product with the FDA

Another processor and exporter of a traditional product into the United States had been exporting the product for many years when the product was held at the Miami port because the FDA Compliance Officer expected the product, which appeared to be thermally processed, to have been filed with the Low Acid Canned Foods (LACF) division of the FDA's Center for Food Safety and Applied Nutrition (CFSAN). The summary of the case for the product, mango kuchela, a traditional Southern Caribbean condiment made from mango (*M. indica*) is outlined in Table 2.2.

The first issue was to establish for the Compliance Officer that the product had been properly assessed by a competent authority, in this case, a Process Authority accepted by the FDA, and was safe for entry into commercial trade in the United States. Another was to confirm that the product had been appropriately filed with the relevant section of the FDA (in this case, the LACF division of CFSAN) or, if not required to do so, that this could be proven. It should be noted that this case occurred before the FDA upgrade of their process filing platform for Low Acid and Acidified Foods to allow for "voluntary" filing of products deemed to be exempt under existing regulations. The approach taken was as follows:

1. The Compliance Officer was contacted and made aware of the assessment of the product by a Process Authority and that the product was shelf stable and therefore exempt for filing.
2. A timeline was agreed by which this information would be made available to him to aid in his decision-making process.
3. As agreed, the detailed information, including information taken from the FDA CFSAN's own website was provided to the Officer.
4. The product was assessed and subsequently released. Shipments are now sent along with the relevant information or this is provided in advance to the importer to be kept on file as the product, not having been filed, does not have a Scheduled Process Identification (SID) Number.

The product was deemed to be exempt because of its general composition, shown in Table 2.3 which shows that the major component of the product, mango (*M. indica*), is an acid food. Further, the product did not have any major constituent that would change the acid nature of the food and the manufacturing process showed that neither was the product heated, nor did it undergo any substantial transformation that would reduce its acidic nature and hence its safety. The manufacturing process was summarized as follows:

- The green mangoes are prepared, sorted, and soaked in chlorinated water.
- They are then crushed using a presanitized manual crusher.
- The crushed mangoes are then soaked in salt/sodium benzoate solution for 72 h, prior to being spin-dried and mixed with the other ingredients according to the formulation.

Ingredients of the product are mango, amchar masala, mustard, palm or soybean oil, and peppers (*Capsicum frutescens*), with the mango (71.5%), amchar and peppers (13%, collectively), all acid ingredients, being 84.5% of the total formulation.

Table 2.3 **Composition and pH of a shelf stable traditional product—mango kuchela[a]**

Ingredient	Percentage composition	pH
Mango	71.5	3.4–3.9
Pepper	7.2	4.2–4.5
Amchar masala	—	—
Garlic	—	—
Soybean oil	—	—
Other ingredients	13.2	—

[a]Those measurement and or amount indicated as "—" were withheld for % composition and not done for pH.

The pH, salt content and water activity were as follows:

Finished product pH	3.0
Salt content	6.1%
Water activity (a_w)	0.838 (23.1°C)

The recommended *critical factors* for the product were:

- Finished pH (at 24°C) < 3.2
- Finished product salinity (%) > 6.0
- Water activity (a_w) < 8.4

It was determined that once the directions given were followed and the critical factors met, the product would be commercially sterile and fall into the category of a *formula safe food*. This, in summary, is the process that was filed with LACF and provided to the Compliance Officer, along with the FDA's own assessment of the filing, which they stated in a technical report which read as follows:

REASON for Exemption
The product(s) appear to be an acid food that contain small amounts of low-acid food(s) and have a resultant finished equilibrium pH that does not significantly differ from that of the predominant acid or acid food 21 CFR 114.3(a) (b); This product is a formula safe product and does not require thermal processing. See information provided in this filing;

FINISHED pH - 3;
Thermal processing method - COLD FILL AND HOLD;
Process Method - CONTROL, FORMULATION
Process Source - Technological Solutions Limited;
Process Source date - 02-APR-14
Initial Temp - 78°C;
Scheduled Time - 20 min;
Scheduled Temperature - 78°C

A similar approach can be taken to other such traditional products that are formula safe and can be shown to be so. Exceptions to this are more complicated products such as those that will be discussed later on in this book in Chapter 6 on formula safe foods.

Conclusions

Food industry professional and exporters need to understand the source of potential contamination of the products that they are handling. It is also important to recognize and understand the potentially significant impact of pathogen-mediated foodborne illness outbreaks because of the highly interlinked nature of the modern food supply system. This has led to extensive spread and impact of organisms such as *E. coli* O104H4, *E. coli* O157H7, *L. monocytogenes*, *Salmonella*, and *C. cayetanensis* for a range of products including vegetables and fruits previously not associated with these organisms. Some of these outbreaks led to the loss of life in multiple states and countries, depending on where the product was distributed. To mitigate this risk and also to comply with market access requirements, effective food safety systems need to be developed and implemented in food handling and processing operations. A HACCP-based food safety or a FSQS compliant with or accepted by the GFSI is now being increasingly demanded by buyers in developed country markets and acknowledged by regulators as a good step toward complying with their requirements. Nevertheless, all countries have specific labeling, ingredient prohibition, and product registration and filing regulations, among others that firms must comply with if their goods are to be allowed across borders. This obliges the food industry professionals supporting the export/import enterprise to become aware of, and keep abreast with the changing requirements of the marketplace as was exemplified through the case studies in the latter part of this chapter. This will continue to be the case as regulators seek to assure food safety for their consumers.

Endnotes

[a]United States Agency for International Development.
[b]Technological Solutions Limited (TSL)'s ISO 17025 accredited laboratory. This laboratory routinely examines a range of foods from multiple countries for extraneous matter, among other things.
[c]Reference to TSL's laboratory, an ISO 17025 accredited laboratory in the Caribbean that undertakes a range of food analyses, including extraneous matter.
[d]"We" refers to Technological Solutions Limited, TSL, the author's company.
[e]Uremia is characterized by acute kidney failure.
[f]Low platelet count.
[g]Dialysis is the process by which waste, excess water, and salt are removed from the blood by a machine that is a replacement for the kidneys which no longer undertake this activity because of permanent damage (i.e., dialysis is a replacement for lost kidney function).
[h]Sporozoites are the infective form of the parasite.
[i]Under the MPE, export growers must comply with a detailed food safety program of practices and pass frequent local inspections by the Integral Program for Agricultural and Environmental Protection, a Guatemalan public–private organization. They must also undergo FDA inspections. A traceability system supported by individual codes applied to each container of raspberries, allows traceback to each individual grower. This allows the export authority of any firm found to be the cause of a food safety problem to be revoked.
[j]The website for the Global Food Safety Initiative is www.mygfsi.com
[k]The Jamaica Ackee Task Force (JATF) was led by the author on behalf of the private sector interests and growers and planned and coordinated the national program to transform the sector.
[l]As described in detail in Gordon, (2015a), this program was implemented by his firm, Technological Solutions Limited (TSL) on behalf of the Jamaica Exporters Association (JEA) and, subsequently, the Jamaica Agroprocessors Association (JAPA), supported by the Government of Jamaica (GOJ), the USDA, the FDA, and the US Embassy in Jamaica.
[m]The Federal Register is available online at https://www.federalregister.gov/.

[n]Available at https://www.accessdata.fda.gov/scripts/cdrh/cfdocs/cfcfr/CFRSearch. cfm?CFRPart=101&showFR=1.

[o]Available at http://www.fda.gov/forindustry/coloradditives/coloradditiveinventories/ucm115641. htm.

[p]Available at http://www.accessdata.fda.gov/scripts/cdrh/cfdocs/cfcfr/CFRSearch. cfm?CFRPart=82

[q]Gray's is a pioneer in the typical Caribbean red and yellow pepper sauces made from West Indian red peppers and red and yellow Habanero peppers. All of its products are naturally fermented and filled in-plant. No dyes or other nonnatural ingredients are added.

[r]Carcinogenic compounds can cause cancer while teratogenic compounds have been shown to cause malformation and other genetic abnormalities in fetuses.

Case study: Improving the quality and viability of a traditional beverage—Irish Moss

A. Gordon
Technological Solutions Limited, Kingston, Jamaica

Chapter Outline

Introduction

Background to Irish moss

Irish moss is both a popular traditional milk-based beverage in Scotland, Ireland, South America, and the Caribbean as well as the naturally occurring and also widely cultivated seaweeds *Chondrus crispus* and *Gracilaria* sp. *C crispus* is commonly called Irish moss or carrageenan moss in parts of Europe and is a protist that, while typically red in color, also varies from green to dark purple-brown (Den Hartog, 1978). Comprised mainly of carrageenan (55%) as well as 10% protein, 15% minerals, and significant levels of sulfur and iodine (USDA Nutrient Database, 2016), *C. crispus* is consumed not only as the beverage, Irish moss, but also as part of deserts and other dishes throughout Asia where it is cultivated along with other seaweeds (Davidson, 2004). Commonly used with agar derived from *Gracilaria*, an agarophyte red algae from the family Rhodophyta, these seaweeds are also cultivated and consumed in Africa, Europe, Oceania, and South America as beverages and jelly and, in Asia, as part of their traditional diet (Davidson, 2004; Goreau and Trench, 2013).

Also known as "seamoss" in countries in the Organization of Eastern Caribbean States (OECS) part of the Caribbean region, Irish moss as a beverage in most parts of the world is boiled with milk before sugar and spices such as cinnamon and vanilla are added. As a traditional beverage in Scotland and Ireland, brandy or whiskey may be added, while in Jamaica, Trinidad and Tobago, and the rest of the Caribbean, it is boiled with milk, cinnamon, vanilla, and other additives to create a thick beverage which is regarded as an aphrodisiac (Mitchell, 2011; Smith et al., 2012). While the traditional beverage can be made from both genera of seaweeds, in the Caribbean region Irish moss or seamoss is typically made from *Gracilaria* spp. Much of the seaweed is found off the coast of several countries throughout the region with St. Lucia being a major source in the Eastern Caribbean in the late 1990s (Smith et al., 2012) when there was an active program to cultivate *Gracilaria* there. In Jamaica where most of the commercially produced Irish moss beverages are manufactured, the seaweed is imported from the Philippines. In Dominica where the majority of the commercial production was located, the seaweed was imported from neighboring St. Lucia, with *Gracilaria debilis* being the preferred source and main import but this was sometimes supplemented by *Gracilaria domingensis* which was grown/sourced with it and sometimes indistinguishable from *G. debilis*.

Profile of a small Irish Moss producer: Benjo Seamoss Agro Processing Limited

Benjo Seamoss and Agro Processing Limited (Benjo) is a limited liability company that was located at Ravine Coque in Roseau in the Commonwealth of Dominica in 1999, at the time of the initial intervention described as part of this case study. The firm produced a variety of traditional beverages using Irish moss (also more commonly called seamoss in that region of the world). The company manufactured the products from marine algae, particularly the agar-rich variety *G. debilis*, previously sourced from St. Lucia but supplies of which were then (at the time) being obtained from Grenada, both nearby Organization of Eastern Caribbean States (OECS) countries. The Irish moss used was a completely natural product which was not cultivated but harvested from the wild. The company was marketing its product locally in Dominica,[a] in several English speaking Caribbean countries and in other Caribbean countries such as St. Maarten at the time. The product had met with reasonable success and had great potential for increased volumes, particularly since beverage products in this category had shown consistent growth globally over many years. In addition, there was a growing market for seaweed products which was estimated to be expanding at a rate of 10% per annum.[b] As a result, the firm's production

Figure 3.1 Benjo Seamoss's original processing plant at Ravine Coque (Dominica).

and exports to the OECS region was expected to increase by over 250% over the next 6 years (to 2005), with exports into EU and US territories and other countries in the wider region being targeted, thereafter.

The company manufactured its products on a single production line on which it produced different variants of seamoss drinks, including seamoss with Bois Bande,[c] plain seamoss (natural, nothing added), and seamoss with milk. Sixty percent of consumers in the regional market for its product prefer seamoss with milk, 20% the natural product, and 8% the seamoss with Bois Bande. Benjo operated from a small facility (Fig. 3.1) with rudimentary equipment using its existing manufacturing protocols. The company, however, had plans for expansion to take advantage of opportunities identified in the market place, as outlined later and had, in addition, realized that as its volumes grew, so did the cost of any quality problems that arose from time to time or noncompliance with market requirements.

In 1999, Benjo was in the second year of implementing a strategic marketing plan which was a critical part of its business plan and which called for a significant increase in sales to its targeted markets over the next few years. Local sales were expected to grow by an average of 10% per annum for the next 3 years, after which the annual increase was targeted at 5%. Regionally, the plan called for expansion in the volumes of its sales by 20% over the next 2 years, then 45% per annum over the following 6 years. Internationally, a 50% per annum growth was expected over the next 5 years, followed by a 35% per annum growth in the succeeding 6 years. All of this would require substantial improvements in production capacity, logistics and sales and marketing capabilities and, critically, effective technical support to facilitate the increases in production, compliance with requirements and access to markets that were being targeted under its business plan. In essence then, the firm could only achieve these objectives if it had strong, effective technical support to address its quality systems, food safety, and food science needs. This upfront recognition of the role that food science and quality systems play in the successful growth and expansion of food production enterprises formed the basis for this case in which support was provided to help the firm successfully implement its business and market expansion plan. It also highlights the importance of considering the technical aspects of a business in developing successful business plans for food production and handling businesses.

Benjo was desirous of modernizing its facilities, addressing existing constraints to production and growth and implementing a recognized Good Manufacturing Practice (GMP) program that would allow its products access into regional and extraregional markets. To make its plans a reality, Benjo realized that it would need technical support. At the time (in 1999), HACCP[d]-based food safety systems were becoming in vogue in Canada and the Australasia region[e] in the dairy and meat processing industries, were mandatory for some categories of products in the European Union, and were required for meats and seafood in the United States of America. However, the HACCP concept had not yet become the norm in the food industry, GMPs being regarded as a good and more practical starting point for most food and beverage production firms trading in developed country markets, hence the interest of Benjo in GMP systems implementation. Benjo's management, advised by its supporting network of partners therefore sought the assistance of a firm that specialized in providing technical assistance in food safety and quality systems with knowledge of the nuances of the Caribbean region, which could assist them in achieving compliance with the requirements for certification to a recognized GMP standard for export to North America and Europe.[f] This formed the basis of a series of technical assistance interventions which were maintained, as required, over an extended period of time.

Products and production processes

At the time of the initial intervention in 1999, Benjo had a total staff compliment of 12, including the owner and manager of the business. The firm operated with the most basic, but scale-appropriate equipment that it used to successfully manufacture products at its then current level of production. This was done according to manufacturing protocols and procedures that had been developed by the owners, as for many other start-up food businesses, largely through trial and error. The beverages were bottled in 5-oz. bottles such as the one shown in the middle in Fig. 3.2, then normally used for carbonated beverages which were quite popular in the region. While initially manufacturing only three variants, Benjo subsequently expanded its range between 2002 and 2006 to include other variants that were also associated with various positive health outcomes, including Seamoss with Oats and Barley and Seamoss with Linseed (Table 3.1).

The original process of manufacturing Benjo seamoss beverages was discontinuous and is outlined in Fig. 3.3 and also shown (selected stages) in Figs. 3.4 and 3.5. The product was made as outlined in the following:

Figure 3.2 Some of Benjo's products in the early 2000s in the original and newer bottles.

Table 3.1 **Selected products made at Benjo Seamoss over the period 1999–2010**

Product type (original)	List of ingredients
Natural Seamoss	Seamoss, sugar, citric acid, sodium citrate, stabilizer, cinnamon, nutmeg, natural flavors
Seamoss with milk	Seamoss, sugar, milk, citric acid, sodium citrate, stabilizer, spices, natural flavors
Seamoss with Bois Bande	Seamoss, sugar, Bois Bande, citric acid, sodium citrate, stabilizer, spices, natural flavors
Product type (new flavors)	**List of ingredients**
Seamoss with oats and barley	Seamoss, sugar, milk, oats, barley, citric acid, sodium citrate, stabilizer, spices, natural flavors
Seamoss with linseed	Seamoss, sugar, milk, linseed oil,[a] citric acid, sodium citrate, stabilizer, spices, natural flavors

[a]Linseed oil is the oil expressed from flax plant (*Linum usitatissimum*). It is thought to have health promoting properties, being rich in the $\Omega 3$ fatty acid, α-linolenic acid, which creates a demand for products containing the oil in the Latin American region and elsewhere.

1. The seamoss was selected, examined and physically washed 3 times (Fig. 3.4A).
2. It was then placed in a steam-jacketed kettle and heated for 30 min or until it softened.
3. The product was then mixed, acidified, and homogenized.
4. The homogenate was held overnight and then water and other ingredients were added.
5. The homogenate was blended (20 min) while being heated and other ingredients being added (Fig. 3.4B).

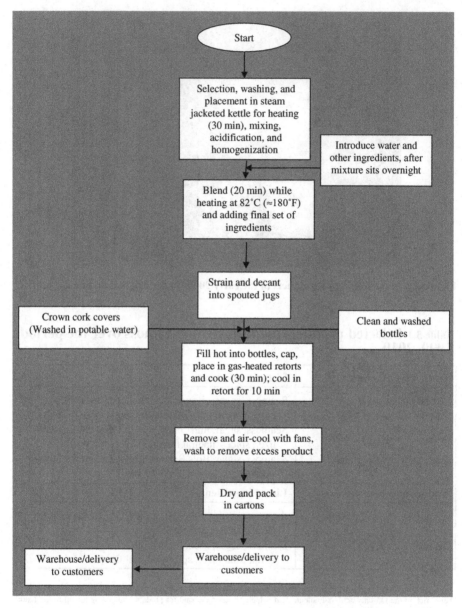

Figure 3.3 Process flow for making Seamoss (Irish Moss) at Benjo in 1999.

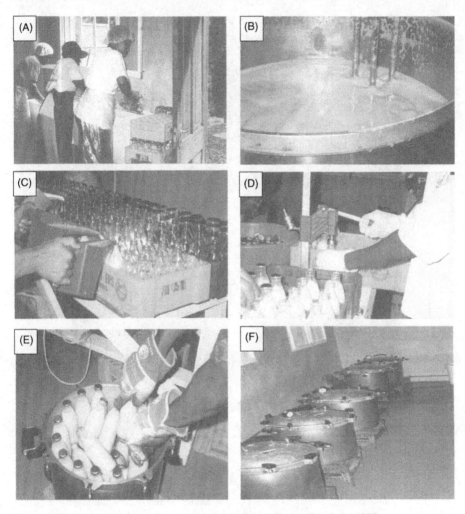

Figure 3.4 Original process for making Irish (Sea) Moss at Benjo in 1999.

6. The finished product was then strained and decanted into spouted jugs, which were used to fill the bottles (Fig. 3.4C).
7. The bottles were capped (Fig. 3.4D) and placed in small "retorts" (pressure cooker units) and heated on gas stoves for about 30 min (Fig. 3.4F) before the pressure cookers were allowed to cool.
8. The pressure cookers were opened and the bottles cooled in the retorts for 5–10 min, then removed and put to cool under fans, washed to remove excess product from the bottles, before being dried, placed into soft drink trays for warehousing and subsequent delivery to market (Fig. 3.5).

This process and production system was the one based on which Benjo had built its business and with which it sought assistance to improve its operations as it sought to successfully pursue its business plan and expand production and sales in third country markets.

Figure 3.5 Cooling bottles of Benjo Seamoss using fans.

Background to the case study

In pursuing its development program in June 1999, Benjo Seamoss and Agro Processing Limited (Benjo) sought the assistance of the National Development Corporation
of Dominica (NDC) and the Economic Development and Agricultural Diversification
Unit (EDADU) of the OECS to provide assistance with addressing technical challenges with its operations and to support its expansion thrust. The NDC supported
the development of Benjo's business plan,[g] which called for the implementation of
a strategy to upgrade the labels for the product, establish overseas distributorships
and to enter the US market through the U.S. Virgin Islands. To accomplish all of this,

the company would need to ensure that it is able to produce consistently high quality product, with good shelf life, packaged in a manner to facilitate easy transportation and distribution. The company had been experiencing sporadic quality problems over the last few years and was hesitant to significantly expand its production without having solved this problem. In addition, it needed to ensure that the product being targeted for its expanded market would be able to meet the market requirements in terms of shelf life, packaging format, and appearance.

Benjo, in conjunction with the NDC and EDADU, therefore requested assistance with a technical diagnostic of Benjo's operations, the objective being to provide a basis for addressing existing constraints to growth, correcting current sporadic spoilage problems and providing a basis for future expansion. In addition, nutritional analyses and shelf life studies were also to be done on Benjo's products. A firm, Technological Solutions Limited (TSL), was contracted with the assistance of the Centre for the Development of Industry (CDI), Brussels,[h] to undertake this initial assignment. This led to the start of a 15-year relationship with Benjo and its product which is the subject of this case study.

TSL's lead specialist[i] visited Dominica and conducted the technical assessment over a 2-day period. The outcome of the assessment and interventions were encapsulated in a report that addressed the operations of the company and its technical and operational strengths and weaknesses at the time of the diagnostic. It also summarized the overall outcome of the interventions and made recommendations as to approaches that the company should take to achieve its goal of addressing existing quality, efficiency, and supply constraint issues, as well as achieve wider market penetration in the Caribbean and beyond. This was followed by a second intervention in October 1999 in which the implementation of the recommendations was the focus. This subsequent visit was to assess the improvements made, to further assist with the implementation of production and quality improvements and to train the staff in the principles of basic sanitation and hygiene as related to the production of seamoss. The assistance given to Benjo in 1999 marked the start of a process of working with this firm to first address existing quality and other needs and then to help the company to grow its business, expand into export markets and significantly improve it product, processing, efficiencies, and competitiveness through systems implementation. At the end of the process of assistance, the firm has addressed several problems, issues, and constraints that also typically affect other small food processors in developing countries, whether they produce low acid canned foods like Benjo does, or other products. This case study presents detailed information on various issues successfully addressed and describes how the firm was able to upgrade its operations to now become an increasingly important player in the Caribbean regional and subregional export market.

For this case study, the author will describe in detail the methodology that was developed and applied in this, and many subsequent cases in which technological solutions were delivered to food production and handling firms to address their current and future business needs. The approach, which was honed during the work described in this case study and subsequently, was further improved and constantly updated as new information and approaches have become available in the various disciplines of food science. It is based on the systematic application of multiple technologies and approaches to address problems or create solutions (whether these are new or improved products, reduced costs, greater competitiveness, new technologies, or simply improving competence of the persons involved through effective training). In this case study, thermal processing, microbiology, ingredient technology, food analysis, process engineering and logistics, and accelerated shelf life determination based on kinetic modeling were among the aspects of food science used. The issues explored and discussed in this case study should prove beneficial for firms, supporting organizations, and professionals in the food industry that have to deal with similar challenges.

Approach to the technical support intervention

For Benjo to effectively implement its business plan, it would need to address and ensure the delivery of consistently high quality product that also met the market access requirements and specific needs of the markets being targeted. The initial technical support intervention for Benjo was therefore divided into two phases. In phase one, it was expected that immediate challenges with production would be addressed through the application of food science to deliver improvements in operations systems, efficiency, and quality. The details of the specific technical assistance to be provided, while outlined in general, would be advised by the outcome of a Technical Diagnostic to be done as part of the first intervention.[j] The second phase of the intervention would assess the effectiveness of the implementation of the solutions presented from the first intervention and also address specific areas of the firms' technical development that would be required for it to apply industry best practices and to become sustainably competitive in the production of seamoss-based beverages. The second phase of the intervention was also expected to build on the first and enable Benjo to be able to successfully implement its business plan.

Phase I: the first set of interventions

The interventions at this phase of the project were designed to address the technical considerations that were known and those that arose as a result of the assessments undertaken as part of the technical diagnostic requested. The interventions at this stage of Benjo's development, among other things, sought to:

- provide the requisite technical support to solve the current spoilage problem being experienced by the plant on a sporadic basis;
- review the existing processing methods and identify ways to reduce losses, improve general quality, and food safety;
- identify the areas and actions needed to improve the operations, in general;
- identify opportunities for efficiency improvements through new processing methods and improvement to processing plant facilities, where appropriate;
- address the needs of the firm regarding preparation for higher volume production and export market readiness;
- identify what would be required, plan, and undertake accelerated shelf life determination on the three variants of Irish (sea) moss produced by the firm; and

- arrange for samples for and undertake the requisite nutritional analyses on each variant of Benjo Seamoss to support the label upgrade being done with a FDA-compliant nutrition labeling declaration.

This phase of the project was therefore expected not only to provide immediate solutions to existing issues, but also to identify and delineate specifically what the firm would need to do to address any of the issues it currently faced should they arise in the future. It was also expected to outline what would need to be done to improve the efficiency of the operations in the short term, while preparing for future expansion. At the completion of the first phase of the intervention, Benjo was to be given a workable solution to their on-going quality problems, a clear direction as to what could be done immediately to improve operations and what changes, if any, were required over time to meet their stated objectives.

The second phase of the intervention was a follow-up to the first visit. Specifically, this phase was expected to:

- assess the changes made by the company and guide any others required;
- train staff in sanitation and hygiene and basic quality issues;
- assess what further technical and other support might be required if Benjo was to achieve its goals;
- examine the proposed site for Benjo's new factory (this arose from a recognition of the constraints of the operations at the time, many of which were imposed by the site at which production was being done) and provide guidance with the technical considerations in moving production to a new site; and
- present an action plan for the company outlining what is required to allow them to address any gaps and/or problems with their existing product(s), systems and/or processes and meet the technical requirements to successfully implement their business plan.

In summary, this intervention was expected to identify any gaps in the existing physical and support infrastructure, systems and practices at the plant (wherever located) specifically related to this product and production line. In particular, specific attention was to be paid to:

1. any factor that was affecting the quality of the product, leading to spoilage;
2. any processing conditions that affect product safety;
3. address efficiency issues by identifying constraints with, and make recommendations to correct:
   a. the product and raw material flow;
   b. the efficiency of the production process from raw material through to finished product;
   c. the generation of waste and losses and opportunities to reduce these;
   d. opportunities for yield improvement or cost reduction;
   e. establishing the true production capacity of the plant at current levels of operation;
   f. technical support requirements;
   g. any staff development requirements; and
   h. requirements for specific equipment and capabilities to facilitate more efficient, profitable production of a safe, high quality product.

Overall, the second intervention was expected to provide the basis for sustained growth and profitability for the company through a food science–based technical strengthening of its operations in the production of Irish moss (seamoss) beverages targeted at a growing regional and extraregional market.

Phase II: subsequent interventions

Subsequent to Phase I and II of this technical support program of assistance to Benjo, the technical support service provider[k] remained engaged, helping them to upgrade their operations and address technical challenges that arose, as requested by the seamoss manufacturer. This involved assistance with new equipment and production facilities, thermal process development and filing, improving formulations, reducing costs and improving competitiveness, as well as training and capacity building for Benjo's staff. TSL also has continued this support as the business transitioned to new ownership as it continues to grow and expand. This subsequent support built on what was done before and highlighted the reality that Benjo Seamoss's success was only possible because of the foundation that had been laid during the 1999 interventions.

The outcome of the interventions in Phase I (first and second intervention in 1999) and Phase II (all subsequent interventions) are discussed in detail in the following sections. The focus is on what the issues were and how these were characterized in the context of food science–based phenomena, how these issues were converted to technical, actionable criteria, and then addressed by the application of appropriate technology to provide a long-term solution. The findings of each aspect of any assessments done is described in detail, as is the characterization of the problem/issue and the derivation and application of solutions. The idea here is to be sufficiently detailed to allow the foundation laid here to be used in subsequent chapters, without having to go into the same level of detail. This will be evident in the reporting on some of the findings of the technical diagnostic. This approach of initially going into detail, followed by assimilation of the information and approach for subsequent, more efficient use, mirrors the actual approach taken by the technical support provider (TSL) in building out its technical support capability and programs for beneficiary firms. It is an approach that is highly recommended for professionals applying food science and technology to building competitive food production businesses.

Outcome of the first intervention

The existing process, layout, and operations

Production operations and existing layout

Benjo's production operations and layout were assessed as part of the first intervention. The plant stored its raw materials in a location that was not connected or adjacent to the production area. These were then moved to the production area prior to processing. All of the production operations occurred in the processing area (Figs. 3.1 and 3.4), with finished product being warehoused in a finished goods storage area that was adjacent to the production area. The production operations were discontinuous in nature, with a gel first being made from the raw seamoss (Irish moss) prior to blending to form the uncooked beverage (Fig. 3.4A–B). This was then filled into bottles (Fig. 3.4C), the bottles capped (Fig. 3.4D) and retorted (Fig. 3.4E) and then air-cooled with fans (Fig. 3.5) and washed to remove any residual drink from the exterior of the bottles and air-dried prior to packing into cartons for storage in the finished goods warehouse.

The processing plant was fairly well laid out and was neatly organized. However, there was significant cross traffic in the plant, this being one of the area of the operations that required improvement. The layout of the existing facility also did not lend itself to the proper implementation of a HACCP-based food safety system which had been identified by TSL as the system of choice for food handling firms at that time. Wet and dry areas were not sufficiently well separated. In addition, while the floor was in good condition and was kept clean using a squeegee,[1] it was not properly sloped and did not have the required drains to facilitate ease of effective cleaning. Despite these shortcomings, it appeared to be possible to rearrange the operations to get better, if not ideal, separation of the sensitive and less sensitive unit operations[m] to achieve more consistency in quality and greater production efficiencies. Further details of the initial assessment as part of the technical diagnostic are given as follow, the recommendations that were made to the firm to address the issues identified or to enable improvements being highlighted in *italics* for ease of reference. In reading through, it may be useful to remember that this intervention was done at a time well before what is now accepted as the norm for GMPs was in vogue in many production facilities throughout the world, including developed countries.

Facilities

The facilities at Benjo were well kept, clean and housekeeping was generally good. There were, however, some opportunities for improvement in the operations noted during the assessment. These are highlighted in the following, along with the pertinent observations.

1. The existing changing room/bathroom facilities both had working sanitary facilities and adequate availability of sanitary supplies at the time of the visit. However, the facilities were very cramped, there was not sufficient space for staff to change comfortably, and males and females had to use the same restricted space.

   The company needs larger, better changing room facilities with hands free washbasins in the staff bathrooms. In addition, at least one hands free hand wash station is needed on the production floor by the entrance from the warehousing area.

2. The lights in the processing area were not covered in such a manner as to prevent possible contamination of the product if they were to break.

   The lights in the processing area need to be protected with a shatter-proof diffuser/tubes with the appropriate ends to prevent a risk of glass contamination of products resulting from breakage of a bulb.

3. Wires (power cords for the equipment) were connected to the equipment by hanging from the ceiling (see arrow, Fig. 3.1). These wires are a safety hazard.

   This is not acceptable practice and should be addressed by having the wiring run in such a manner as to not be a hazard to health and safety (e.g., in conduits along the wall).

4. The pest proofing screens by the front of the building were torn and needed replacing.

5. The processing room was too hot, largely because of the method of cooling and the lack of extractor fans.

Addressing both of these will improve the conditions for work and also result in reduced incidences of spoilage.

6. The ingredient and supplies warehouse was musty, not always well organized and was adjudged to possibly represent a mold spoilage threat to the ingredients stored there.

The ingredients and supplies warehouse needed better housekeeping. Better airflow was also needed.

7. Detergent was found stored beside xanthan gum and with other food ingredients. This is not to be allowed.

All chemicals must be stored separately and kept in a locked storage. Chemicals for use in foods should be clearly separated from those used for other purposes.

8. It was observed that packaging material and ingredients were being stored in the same area, although not in close proximity, with chemicals used in various aspects of the operation of the plant.

Cartons, bottles, inserts, and other packaging materials need to be stored separately from ingredients and these also need to be stored in a different place from chemicals. The latter must be kept in a locked storage area.

9. Raw materials were being warehoused in different areas of the facility and also in different buildings.

It would be better to store all similar materials in one place.

10. The existing plant has many ledges that could and did harbor dust which is a contamination hazard. This was not suitable for a food processing facility.

Until there is a change in the facilities where the product is manufactured, it was recommended that Benjo ensured that these ledges were scrupulously cleaned every day.

Food safety and quality assurance practices

The back door and windows leading into the processing area were being left open during the day. Some were not properly pest proofed with appropriate sized screens. This was identified as an inappropriate practice as it created the opportunity for pests to enter the factory and possibly end up in the product.

The doors and windows must be kept closed to prevent the entry of insects and other pests into the production facilities.

Personnel and practices

1. As indicated previously, the doors and windows were being left open. Largely because of the heat in the production area due to the way the product was being cooled (Fig. 3.5).

The problem with the heat could be addressed by procuring large extraction fans and placing these over the windows at one end of the plant. This can be used to pull hot air consistently out of the processing area and address the problem with heat.

2. It was noted that some staff wore inappropriate footwear in the facility, including open footwear (slippers) and other nonsafety shoes.

   This was identified as an unacceptable practice that must be addressed, with all employees being issued with boots or other acceptable footwear to be used on the production floor.

3. Several members of staff were in clothes worn from their homes to work. This was flagged as unacceptable for a food production environment.

   The firm was advised that all persons entering or working in the production areas needed to be properly attired with hair covering and overcoats/coverall that did not contain buttons and did not have pockets above the waist.

4. There was evidence of low-level pest infestation in the delivery warehouse (next to the production area).

   Adequate pest control procedures needed to be implemented. This would involve getting in a competent pest control company or developing and implementing a rigorously monitored pest control program (in-house).

5. Members of staff were seen wearing jewelry (earrings and watches) in the plant.
   This is prohibited.

6. Detailed work instructions were not available or documented. However, posters reminding staff of appropriate steps in the production process were placed at various points in the production area.

   This was noted as a good practice. Nevertheless, it was pointed out that work instructions needed to be documented for all of the steps of the process to ensure standardization of the output and allow for flexibility in the manufacturing operations.

7. Benjo's production staff demonstrated practices in some instances that were suboptimal because they had never been exposed to training in the proper way to undertake the processing being done.

   It was recommended that the production and all other staff that may enter the production area undergo the appropriate training. This would include Basic Food Processing, Sanitation and Hygiene, Hazard Analysis Critical Control Points (HACCP) and Quality Assurance for all members of staff. Selected persons could be targeted for higher level training, including supervisors who should be exposed to a residential program tailored to suit the specific needs for the seamoss processing/beverage industry.

Raw materials handling

1. It was observed that some bags of ingredients were left open after being used for dispensing material for each batch.[n]

   This is not a good practice and can result in contamination of the product with extraneous (foreign) matter. Benjo was advised that this practice must be discontinued. All containers/packages must be resealed as soon the material required has been removed from them.

2. It was observed that the firm was using recycled bottles as one source of bottles in which the product was being packaged. It was also noted that the system of handling the recycled bottles could result in an increase in the risk of spoilage.

The practice therefore, while commendable, could become a source of intermittent spoilage problems if the bottles were not effectively cleaned and sanitized. It was recommended that bottles to be recycled should be washed immediately on receipt and cycled through a sanitizer bath with verifiable levels of antimicrobial cleaners and sanitizers.

3. There was no evidence that the water used in the plant was chlorinated or, if so, that the level of chlorination was known.

The level of residual chlorine in the water being used for washing bottles and for other purposes in the plant needed to be determined on a daily basis and adjusted to meet an agreed minimum residual chlorine target.

4. There was no system in place for tracing ingredients and inputs through to products and products through the distribution system to the final retail outlets.

It was indicated that each batch of products needed a unique batch number that would be used to identify it in any market and allow for traceability back to the time processed and each ingredient used. Each ingredient also needed to be assigned a unique trace code to track its use in a particular batch of products. These were outlined as absolute requirements if serious exports into many markets were to be pursued by Benjo. Further, it was recommended that for the ingredients, it may be possible to use their manufacturer's batch numbers as their trace codes. If none existed, as in the case of sugar (which was uncoded in many instances at that time) the receival date or other firm-assigned coding could be used as a trace code.

Documentation

Based on the assessment done, the firm was deficient in several areas where documentation was required.[o] The following were the documentation that needed to be developed.

- Detailed ingredient specifications for all raw materials and packaging inputs.
- Detailed in-process specifications.
- Finished product specifications.
- Detailed manufacturing instructions (Standard Operating Procedures).
- Documented cleaning and sanitation procedures for all areas of the operations.
- Sanitation Standard Operating Procedures (SSOPs) to support the sanitation program.
- Written quality assurance procedures, including the chemical and microbiological tests required.
- A traceability and recall procedure.
- A pest control program.
- A preventative maintenance program.
- A detailed training program for each member of staff and/or to develop a desired competence for the company.[p]
- A program covering GMPs requirements.

Processing

The assessment was able to determine that despite the findings noted, it is evident that there had been significant improvements in process control during the manufacture of Benjo seamoss, especially after the return of the general manager and production supervisor from the thermal process training undertaken the previous year. This was used as an example of how targeted training in areas that build specific knowledge and production skills can be useful to a Micro, Small or Medium Sized Enterprise (MSME) involved in food production and how what was learned was immediately put to use in improving the operations. Nevertheless, it was indicated that there were other areas where further improvement was needed. These were as indicated in the following.

1. The process of manufacturing the seamoss drinks as it was being done was too discontinuous. This would restrict the ability of the company to grow its volumes further without adding significant cost (in personnel and equipment).

   A critical examination of the process to have better, more continuous flow and removing the bottlenecks was required. This, however, could not be done without some experimentation to determine how the changes would affect the characteristics of the finished product. This should be done before any decision was taken as to the kind and capacity of new equipment to be bought to assist in creating room for expansion of the company.

2. The process as it was currently being done took too much time (as noted previously) and was too complicated with too many steps.

   It needed to be simplified, a process that could be engaged during the implementation of the recommendations coming out of the restructuring of the process to make it more continuous and remove bottlenecks, as mentioned previously.

3. The formulations were all nonscalable and still had the same irregular units of measurement that was used to develop the products initially.

   The formulations needed to be converted to standard format (percentages) to make them scaleable and reduce the risk of errors.[q]

4. The cooling process was far too long and was very likely to create the opportunity for the outgrowth of any spore forming organisms that may have been present in the product or, at the very least, serve as a potentiator of their outgrowth at a later time. This may have been a contributor to some of the quality problems encountered.

   Cooling under a constant flow of chlorinated water was recommended as much preferred, particularly if the water was chilled. This would necessitate building/purchasing a cooling box/spray cooling system. In the interim, an alternative method that involves water cooling should be devised.

5. There was variability in the fill levels and internal temperatures of the product because of how it was being done with long spouted jugs and then being left to sit until a retort was available for further thermal processing. This was also a possible source of some of the sporadic spoilage problems and other quality problems being experienced occasionally.

   This would need to be addressed by automation of the filling process using a simple piston or volumetric filler.

6. The Production Manager used to check off the addition of ingredients to each of the batches to ensure standardization of each batch but the practice had been discontinued.

   It was pointed out that it is important that the actual amount of ingredient added to each batch was recorded to ensure standardization and to allow for traceability all the way back to the blending process in case of problems.

7. In some cases, the product was not being cooked for the stipulated times, nor were they retorted at 15 psi. This was because the temperature of the retorts did not always quite get up to 250°F before process timing started. Another cause was that retorting was stopped before the product had achieved 12 min at 250°F and 15 psi, as required by the process then being used.

It was indicated that it was critical that Benjo established and stuck to a standardized thermal process. Until and unless a new process was established, all batches should be cooked as recommended. The retorts should be allowed to reach 250°F and 15 psi before process timing was started.[r]

Equipment and supplies

The company used equipment that were suited for small-scale processing and did not have the facility for a laboratory to assist with quality assurance but used the services of the Produce Chemists' laboratory. To facilitate growth, there were some areas where changes would be required. These are highlighted as follows.

1. *The firm needed to have a small, practical capability to test process control variables on-site. This included thermometers to measure temperature, at least a handheld refractometer to measure °Brix, and the means to measure pH and acidity of the products (a pH meter and titration apparatus). It was important that the Benjo was able to check at least the pH of any product that appeared to be spoilt and further, they needed to be able to get basic microbiological tests (such as Total Aerobic Plate Counts and coliforms) done on water being used and on product as a part of the company's quality assurance program.*

2. The bottles now being supplied were found to be highly variable in quality and often could not be properly sealed with the crown caps.

   This was identified as a potential source of sporadic quality problems and it was recommended that it should be addressed by dialogue with the supplier or sourcing of an alternate supplier.

3. The facility was using only pan and other analog balances for weighing, creating the opportunity for inaccuracies in weighing that could contribute to variability in the product.

   It was recommended that they needed to get digital balances for weighing to reduce the risk of errors that could contribute to quality issues.

4. The retorts (small pressure cooker units) being used for thermal processing were not only limiting in size but also meant that multiple different processes were possible for each batch of product depending on the control of the process in each unit. This also would contribute to variability in the product within a batch from time to time and in a manner that was not predictable as separate data was not kept on each individual unit, nor was the system of traceability in place (nor would it be practical, given the small size of the batches) to trace individual bottles back to a particular cook.

   The company therefore would need to acquire larger retorts (or a single large unit) that would do much large volumes per cook. This would also necessitate a boiler to produce the steam required, an automatic or semiautomatic filler, an automatic capper with the appropriate turntable to facilitate continuous filling and capping, and pumps of the appropriate type to move the semiprocessed seamoss around from stage to stage, inclusive of filling.

Production capacity

The company produced six batches of product per day operating for 5 days/week. Production of the Irish moss gel took 1.5–3 h, the blending of the gel 20–30 min, retorting 15 min, initial cooling 10 min, secondary and tertiary cooling 120 min, and washing, wiping,[s] and drying the bottles 45 min. The total time taken by the process was approximately 6 h and 40 min (approximately 7 h), that is, just under 1 day to produce bottled product from gel. The overall production time per batch was far too long, impacted by the time required to produce the gel which created a major bottleneck in the process, and the need to leave the gel overnight before continuing production the next day.

As an initial first step to address this relatively inefficient way of producing the beverage the following were recommended:

- *The company decouple gel production from the other aspects of the process, thereby allowing greater productivity than would otherwise be the case.*
- *Productivity could be increased significantly if the time required for the production of the gel was shortened. This would require appropriate R&D work to be done to redesign the process.*
- *On-going production of gel throughout the day would increase throughput if the sequence and timing of the other process operations were amended.*

Based on these recommendations for changing the logistics of the process, and the specific investigation into the sporadic spoilage and other quality issues that arose occasionally (see subsequent sections), the process was amended as shown in Fig. 3.6. This, along with the implementation of the other recommendations, was projected to be able to increase Benjo's production by 30%.

An assessment of the quality-related issues and proposed solutions

The problems being experienced with sporadic spoilage and inconsistency of Benjo's seamoss product were analyzed with a view to determining causes and developing possible solutions. The basic manifestation of the problem was the visible separation of the product and formation of a grayish cloud at the base of the bottle with the product in the middle being transparent. There often would also be evidence of apparent fermentation, with the product tasting bitter and smelling sour. The product would sometimes also be yellow in appearance and there were cases where the product appeared visually acceptable but would be objectionable when ingested.

An assessment of the problem revealed that it occurred most frequently when the production area was particularly hot and congested and when the product took longer to cool than was normal. It was also observed by the Benjo team that, in the past, the incidence of the problem increased during the summer months. The observation (mentioned previously) that product was not always receiving its full thermal process was identified also as a probable contributor to the problem and especially its sporadic

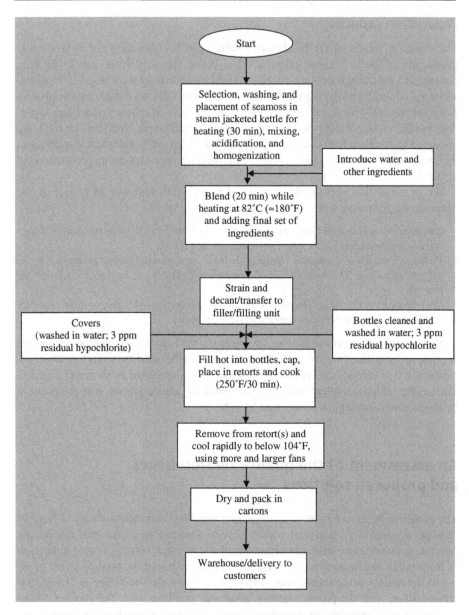

Figure 3.6 Upgraded process flow for Benjo Seamoss after the initial intervention.

and unpredictable nature. These were the general causes. Each is highlighted in the following and the specific solutions recommended are indicated in *italics*.

1. The product was being underprocessed in some cases, as noted previously. This resulted in some units not receiving enough of a thermal process to destroy all of the viable microbial spores present.

All batches of products must get the prescribed cook to ensure optimal destruction of microorganisms.

2. The thermal process being used had not been verified by a process authority as being effective in delivering a commercially sterile product capable of being exported into those countries that require a process to be filed with their regulatory authorities prior to commencement of exports.

   The thermal process being used for the product needs to be verified and filed with the US Food and Drug Administration (FDA) if the product was ever to be exported to a US territory, as was indicated in the company's business plan.

3. The way in which the product was cooled was perhaps one of the major causes of the sporadic spoilage problems.

   The product needed to be cooled down to below 40°C (104°F) as rapidly as is possible. This may be done as described earlier, using a water spray or immersion cooling tunnel. In the interim until this solution could be implemented, it was recommended that more and larger, more power fans be procured to achieve more rapid cooling.

4. Assessment of the vacuums of several bottles of product noted variations. One had a vacuum of 5 mm Hg, another 2.5 mm Hg, while others had no vacuum. A proper vacuum is important to ensure that commercial sterility is maintained in the thermally processed product that is expected to maintain a hermetic[1] seal.

   The variations in the vacuum were likely due to variations in the bottle and caps used. A product such as the seamoss beverage being made ought to use a cap better able to hold a vacuum than the crown-type cap that was being used. Further, even if the cap and bottle could not be changed in the immediate future, there would need to be much better selection of the bottles to be used to ensure adequate vacuums and closures. It was also suggested that the vacuums on representative samples from each batch of products made should also be tested on site, if possible, to give an indication of the sufficiency of the hermetic seal being delivered by the process, per batch.

5. The previous discussion indicated that the packaging of the product was not sufficient to prevent recontamination of the product after thermal processing.

   The packaging needed to be upgraded to use the correct kind of bottle and, possibly, cans. This would require the use of twist-off caps and the corresponding bottles that were sterilizable and compatible with equipment that could facilitate faster production times, such as semiautomatic filling and packaging equipment.

6. It appeared that storage of the product at temperatures that would facilitate the growth of any viable spores that survived processing also played a role in the sporadic spoilage experienced occasionally. This had to be controlled to eliminate it as a potential source of spoilage.

   It was recommended that care should therefore be taken during storage of the product to avoid temperatures above 30°C (86°F) that would promote spore outgrowth and, consequently, spoilage of the products.

7. Some of the variations in product appearance and texture noted appeared not to be due to commercial nonsterility, but rather to inappropriate viscosity and texture (rheology) of the product. This was thought to be due to variations in the quality of the raw material supplied, in particular the seamoss. An examination of the raw material in storage, records of raw

material purchase and sourcing confirmed apparent differences in the seaweed being used in the product. While the nature of seaweed is that there would be occasional variability in the content of carrageenan and agar, the major gelling agents naturally present, differences in species or the use of completely different genera (e.g., *C. crispus* vs. *Gracilaria* sp.) would inevitably lead to different processing outcomes. This is because the composition of each of these seaweeds is completely different in relation to the percentage of agar versus carrageenan that is present, with *Chondrus* sp. being mainly carrageenan and *Gracilaria* being mainly agar (Den Hartog, 1978). While understanding that there were differences, Benjo was not aware of the nature of the differences and the impact it could have on the variability of the product. This appeared to be the cause of some of the variation in quality noted.

8. Benjo was sourcing seaweed (seamoss) from St. Lucia and getting a mixture of *Gracilaria* sp., *G. debilis* and *G. domingensis*, both of which contained varying amounts of gelling agents. In addition, occasionally the seaweed being supplied was neither, but likely *C. crispus* which was also available from the same source. This contributed to the variability in the rheology of the product that occasionally occurred and, because of the unavailability of traceability and record keeping systems that tracked the type of seaweed used, it was not possible to specifically identify which type of seaweed delivered what kind of sensory properties. This could be controlled by specifying more carefully the type of seamoss required and also possibly doing analyses to properly quantify the characteristics of the seamoss that is best suited to the production of a consistently acceptable product.

It was recommended that Benjo develop and buy seaweed based on specification which they should issue to their suppliers. It was also recommended that they determine which type of seaweed produced the kind of product that was desirable and what impact each different type had on the product, so that they could make adjustments should the preferred input not be available.

The recommendations aforementioned, when implemented were expected to significantly improve the company's operations, reduce or eliminate the quality issues being experienced and improve the production capacity of the plant. It was recognized that Benjo would not likely be able to do all that was recommended in the immediate future, hence the recommendation of temporary measures to address the quality issues and improve practices.

During the second phase of the intervention, the progress made with each of the areas identified in the first phase were assessed to determine how effectively they had been implemented and whether further modification was needed. In addition, other aspects of technical assistance, including training, shelf life determination and guidance with plant layout was provided as well as assistance with packaging upgrades. The second intervention, its assessment of the impact of the first intervention and the outcome of the specific technical support undertaken under the second phase of the process are detailed as follows.

The second intervention

The second intervention assessed the progress achieved in implementing the recommendations from the first visit, evaluating possible sites for the new factory and reviewing market performance to ensure that the quality problems had been successfully addressed. In addition, the staff of Benjo was trained in Sanitation and Hygiene and production-relevant practices. The objective of the training was to begin to build a sustainable capacity for high quality production at the company. During this visit, other requirements for success were also identified and issues related to shelf life, nutritional labels, and plant layout and location addressed.

Observation of practices and improvements to the process

A review of practices at the plant was undertaken to identify the need for further improvements. The following was noted:

1. The practices in the plant had improved. All persons were wearing hair covering, however members of staff still were not properly attired in company-provided uniforms. These were understood to be on order.
2. Filling was now being done in a far more efficient manner, such that capping had now become the bottleneck to the more efficient completion of the process. It was suggested that the process could be done much more efficiently using a mutlihead filler and in-line capper, such as those available commercially.
3. Based on the recommendation, six new cookers had been purchased and this had increased the rate of production by reducing the constraint of available cooking capacity.
4. The warehouse was much better organized, and the basis for a traceability system had begun to be implemented.
5. In summary, many of the issues raised, including materials transfer and weighing were in the process of being addressed.

Plant location, design, and layout

As noted in the initial intervention, the existing design and layout of Benjo's plant while adequate for the small scale of the start-up operations could be improved to provide greater efficiencies and become more compliant with industry best practices. Any new layout for the existing plant would be dependent on the equipment to be acquired and the changes made to the process in keeping with the recommendations. Alternatively, if a new plant was to be developed, the location and design of that facility would have to be considered before an acceptable layout could be developed.

In the second phase of the intervention, several options for a new plant were evaluated and two potential sites for the new facility were visited. A site was selected based on its ability to comply with the requirements for a modern food production facility and its ability to accommodate the design of a proper layout for the plant. Steps were taken to begin the process of transition to this facility with a layout that TSL provided for Benjo. The new facility with the availability of more space and the ability to separate activities to achieve better control and the production process that eventually ensued there are shown in Fig. 3.7.

Shelf life of seamoss products

The shelf life of three variants of seamoss were determined using accelerated shelf life testing. This process was based on a methodology that had been developed by Brown

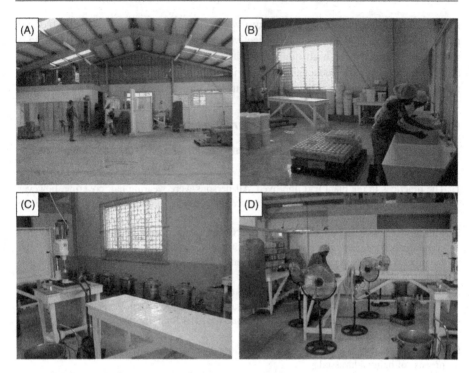

Figure 3.7 Upgraded production of Benjo Seamoss postinitial intervention.

and Gordon (1995) for beverages based on the approach spearheaded by Labuza (1984) involving the application of the principle of the Arrhenius relation (Eq. 3.1) to predicting changes in specific characteristics of the product due to acceleration of their deterioration. The approach involves the application of a stressor to accelerate the rate of deterioration of the product, having evaluated and understood the mechanisms by which the product spoils or becomes unsafe, and then using this information to predict shelf life under normal commercial conditions of handling (Singh and Cadwallader, 2004). The data are then used to determine the index of reaction rate differentials at temperatures differing by a factor of 10, or Q_{10}, which is used to predict the shelf life (θ) of the product. Formally defined, the Q_{10} is the ratio between the reaction rate constants when the temperature of the reactions differs by a factor of 10 (Eq. 3.2). The validity of the approach had been widely demonstrated as many important aspects of product quality deterioration have been shown to follow an Arrhenius relation. The Arrhenius equation is:

$$k = k_0 e^{[(-E_a/R)(1/T - 1/T_0)]}$$

$$(3.1)$$

where k_0 is the rate constant at T_0 (in Kelvin, K), E_a the Arrhenius activation energy (J/mol) and R is the gas constant (J/mol K).

The Q_{10} is determined by the equation:

$$Q_{10} = k_2(T+10)/k_1(T) \tag{3.2}$$

where k_1 is the rate constant at temperature T and k_2 is the rate constant at temperature $T + 10$.

The shelf life (θ) is then calculated by determining the time the product is projected to be acceptable, given the Q_{10} determined as previously. It should be noted that the actual shelf life of products is influenced by a wide variety of factors including packaging, specific product composition, storage, handling and transportation conditions, and other factors. Excellent discussions on the topic of shelf life and accelerated shelf life determination have been provided by Labuza (1982, 1984) and several others (Labuza and Schmidl, 1985; Martins et al., 2008; Singh, 2000; Singh and Cadwallader, 2004).

While storage at various elevated and other predetermined temperatures is usually the mode of acceleration of deterioration used, other modes of failure such as the presence and intensity of light and the availability level of selected gases have also been used.[u] For the determination of the shelf life of the Benjo seamoss drinks, the approach used involved storage of the products at different temperatures (4, 30, and 40°C) and measuring key product attributes on an ongoing basis until deterioration beyond reasonable acceptability was achieved. The characteristics monitored included microbiological indices of quality and safety (total aerobic plate count, aerobic and anaerobic spore counts, *Salmonella* and total coliforms/*Escherichia coli*), pH, free fatty acids (FFAs), and percent sugar (measured as °Brix). Selected data on the characteristics of the Natural Seamoss and Seamoss with Bois Bande are shown in Table 3.2.

An assessment of the three main products over the period showed that the pH of the products correlated most closely with the sensory deterioration that was determined

Table 3.2 Selected data for the accelerated shelf life determination for Benjo Natural Seamoss and Benjo Seamoss with Bois Bande

Characteristic	Natural Seamoss		Seamoss with Bois Bande	
	30°C	40°C	30°C	40°C
Date: 16.08.99[a]				
pH	6.62	6.60	6.21	6.03
Free fatty acids	0.12	0.13	0.11	0.06
Date: 06.12.99[b]				
pH	6.36	6.20	5.60	5.89
Free fatty acids	0.07	0.10	0.14	0.13

[a]This was data for week 1 of the study.
[b]This was data for the final week of the study. The study was done over a 4-month period.

Figure 3.8 Typical Seamoss (Irish Moss) product sold in the Eastern Caribbean.

by sensory measurement throughout the period of shelf life testing. As such, this was the parameter used for the determination of the Q_{10} and for calculation of the predicted shelf life of the product. The outcome of the assessment showed that the products all had a shelf life in excess of 6 months under normal distribution conditions (28°C). Based on kinetic and other calculations using the previous equations, the shelf life for the Seamoss with Bois Bande, the Natural Seamoss, and Seamoss with Milk were 7.5, 7.5, and 6 months at the normal mean distribution temperature (28°C), respectively. At refrigerated temperatures (4°C), the shelf life was determined to be a minimum of 1 year and 3 months for Seamoss with Bois Bande, 1 year and 4 months for Natural Seamoss and 1 year for Seamoss with Milk.

Opportunities for cost reduction/profit improvement

The second intervention with Benjo focused on opportunities for reducing the cost of the product and improving its profitability, important activities that are also often the purview of quality systems practitioners in many companies. For Benjo, there are several opportunities for reducing the cost of its products while maintaining the quality as perceived by the consumer. These, if implemented were expected to significantly improve the profitability and competitiveness of the operations. The specific of these are outlined as follows.

- The use of a less expensive source of seamoss.
- Converting to the use of commercially sourced milk powder rather than the evaporated milk that was then being used in the Seamoss with Milk product. This would have an immediate and significant impact on formulation costs.
- The use of a slow turning massager/blender to wash the seaweed rather than do it by hand as it was being done (Fig. 3.4A). This would improve efficiencies and throughput.

- The use of larger, fixed, steam-driven retorts for thermal processing. This would also improve efficiencies and throughput.
- Filing with a continuous piston or volumetric filler linked to a capper to give continuous capping, that is, a hot fill and seal machine.
- Cooling bottles under a water spray/immersion cooling tunnel (with water recycling), as previously indicated. Alternatively, cooling could be done using a heat exchanger prior to filling. This would require a continuous processing system and there would be the risk of getting the plates clogged if the process were not very carefully controlled.
- Using an automatic carton taping machine to finish cartons instead of doing this manually. Eventually, a palletizer may be used.
- Sourcing and using caps that better fit the current product, thereby reducing vacuum losses and spoilage. It would be even more preferable and highly recommended to move to using twist off caps and lug-closure bottles (as shown in Fig. 3.8) which would deliver a much more consistent and assured hermetic seal.
- Using alternate packaging formats such as cans offer the opportunity of significantly improving throughput but, more importantly, reducing losses due to breakage in the trade. This packaging format also would put the product into a similar category to the leading product of this type in the region, Nestlé's Suppligen, thereby opening up a whole new market that is currently untapped by the product.

In general, as is evident, there were several opportunities to improve efficiencies through automation/semiautomation. These would allow staff to work more efficiently and comfortably while producing a much higher volume of products. These opportunities were automation of washing the raw material, blending, filling the bottles/containers, centrally controlled retorting, cooling of the product, labeling, and packaging.

Summary of new equipment required

The following new equipment were identified as those initially being required to facilitate an expansion in production in the future.

1. blending tanks
2. retorts
3. a steam generator (boiler)
4. an automatic or semiautomatic filler
5. an automatic capper
6. accumulator (turntable) to facilitate continuous filling and capping
7. transfer pumps (2)
8. a labeler
9. air compressor

Future requirements

The future requirements of the company for continued growth and meeting its business objectives were summarized and presented to the firm and its partners as listed here. These exclude those already presented previously.

- Upgrade of quality assurance systems.
- A proper steam-driven retort was needed.

- An appropriate venting schedule and configuration would need to be established for the retort.
- The thermal processes needed to be refined and verified by a process authority and filed with US FDA.
- The opportunities for cost reduction and profit improvement indicated previously should be pursued by the company.
- A HACCP program should be implemented at the plant as soon as was feasible, but as soon as possible after and perhaps coincident with moving to new facilities.[v]
- The company needs to embark on a program of research and development, focusing on new sources of seamoss, ingredient substitution, cost reduction, development of other products and other issues.

Subsequent technical interventions

Several subsequent technical interventions were undertaken over the ensuing years to assist Benjo in implementing the technical aspects of its original business plan. These included movement to newer, larger facilities, equipment, packaging, and processing upgrades, expansion of the firm's markets and general improvement in practices, as recommended. Collectively, these led to improvements in the efficiency of operations and an expansion in Benjo's production capacity by over 80% (30% in the period immediately after implementation of the recommendations from Phase I and a further 40% after Phase II). The improvements in the plant layout and operations on setting up at the new site are shown in Fig. 3.7. The upgraded packaging is evident in Fig. 3.2 (the two bottles with twist-off caps vs. the crown capped bottle in the middle) and was now comparable with other competitor products in the regional market[w] (Fig. 3.8).

This enhanced production allowed the firm to expand its footprint and, eventually, attract the attention of another regional entity that bought the business and moved production to Trinidad. This firm has upgraded the packaging (Fig. 3.9) and continued the technical upgrades recommended and, while not yet going to the enhanced and more efficient type of cooling recommended, has implemented several of the other recommendations, including an improved process flow (Fig. 3.10) and the use of a scale-appropriate, commercial horizontal retort (Fig. 3.11B). The production operations are now semiautomated with many of the controls required to ensure consistently high quality, safe product have been implemented.

Conclusions

The program of technical assistance for Benjo Seamoss and Agro Processing Company Limited was designed to address the firm's food science–based technical support needs through two structured interventions which were expected to deliver a detailed plan for upgrading their operations, over time. This was followed by subsequent targeted technical support to assist Benjo in realizing their business objectives. Collectively,

Figure 3.9 Current fully sleeved packaging for Benjo Seamoss.

these have moved the company to where it is today, with many specific technical issues that are widely applicable to many similar firms globally being successfully addressed. The specific objectives/issues to be addressed and approaches applied are summarized in this concluding section.

The first and second phases of the technical support to the firm anticipated that, at the end of the project, the company would at a minimum, clearly understand what was required in improvements to its quality assurance program to improve product quality and consistency, reduce losses and meet external requirements. The company and its management had this understanding and had improved their quality assurance such that the problems they were previously experiencing had been eliminated. The firm also had begun to implement guidelines for an upgraded quality system and the prerequisites for HACCP. Specific training requirements had been defined to improve the competence and capabilities of their management and staff and the training program has begun to be implemented. Benjo also had a clear idea about what was required to upgrade its operations to improve efficiencies and production capacity, as well as its ability to access new markets. Additionally, new labels and upgraded packaging were in the process of being introduced. As part of this process, nutritional declarations available for its three core products had been delivered. The company also had specific directions as to what to do to reduce costs and had already seen improvements in its production. Finally, Benjo had scientifically sound shelf life for each of its core products delivered through the use of accelerated shelf life determination.

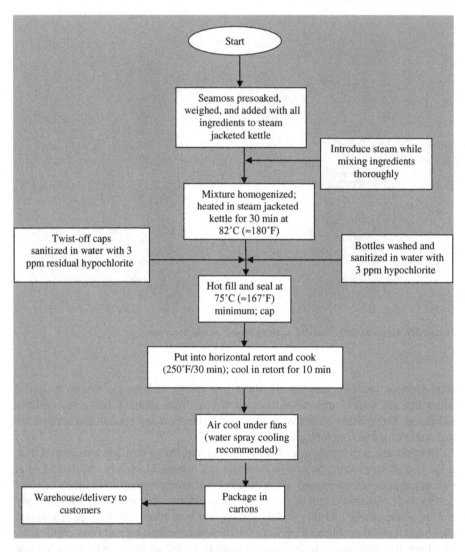

Figure 3.10 Current production process for Benjo Seamoss in new plant.

This case study has shown how a systematic approach to identifying and building a food science–based technical program to improve a firm's food safety, quality, and competitiveness through a planned series of goal-oriented interventions can work, given the reality of small firms, with limited capabilities. In this case, the firm was operating in a developing country environment with its own challenges in terms of access to supplies and resources. Nevertheless, similar process will work well for QA team members, food safety practitioners, and other technical staff of other developing country firms, as well as those in other food manufacturing companies, wherever located. This case study addressed practical steps for dealing with

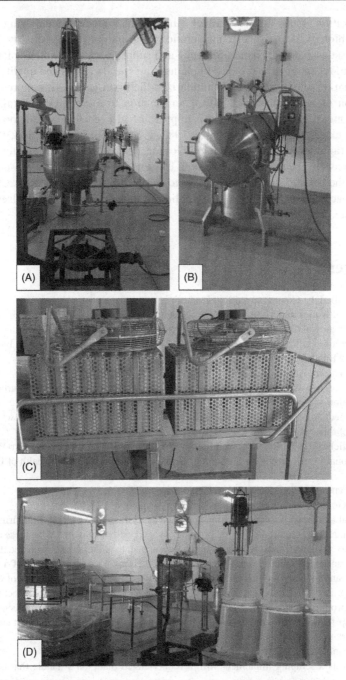

Figure 3.11 New plant and semiautomated process for making Benjo Seamoss.

standardization issues that also affect other SME producers, identifying some of the microbiological and other considerations that may influence quality and food spoilage issues and addressing efficiency and supply issues, as well as plant layout and design. It examined the role of implementing specifications and quality systems to meet market requirements and summarized the first use of TSL's approach to the application of accelerated shelf life determination to a complex beverage matrix like Irish moss. Previous uses of this approach had been for canned foods which have been extensively studied (Labuza, 1982, 1984; Labuza and Schmidl, 1985) and light beverages and juices (Brown and Gordon, 1995), the shelf life determination of which has also been well documented. It is expected that the case presented here will form the basis of success approaches to similar problems by FSQS practitioners. The approaches will be adapted and applied to other cases presented elsewhere in this volume.

Endnotes

[a]Formally called the Commonwealth of Dominica.
[b]Seaweed Cultivation and Marine Ranching, Kanagawa International Fisheries Training Centre, Japan International Cooperation Agency (JICA).
[c]Bois Bande or "hard wood" is a natural product made from the bark of the tree by the same name. Bois Bande's botanical name is *Richeria grandis*.
[d]HACCP—Hazard Analysis Critical Control Points.
[e]Australasia region—mainly Australia and New Zealand and their major trading partners in the Pacific region.
[f]This firm, Technological Solutions Limited (TSL), provided the support to Benjo which forms the basis of this case study.
[g]Eugene, Michael, 1998. Business plan for the expansion of Benjo Seamoss & Agro Processing Ltd., National Development Corporation of Dominica, Roseau, Commonwealth of Dominica, W.I.
[h]Now the Centre for the Development of Enterprise (CDE).
[i]Dr. André Gordon, the author.
[j]A Technical Diagnostic is a combination audit and assessment that evaluates the firm against a range of standards, requirements, and practices as well as the firms stated business objectives and delivers a detailed, step-by-step implementation plan to address the technical deficiencies/noncompliances identified. It examines not only the compliance or otherwise of a firm with regulatory and industry requirements and best practices, but also the suitability of the equipment being used, packaging, practices, and other areas of the operations that affect the firm's ability to produce in a compliant, sustainably competitive manner. It is highly recommended for start-up firms or those seeking to significantly expand their operations, particularly into export markets.
[k]TSL—Technological Solutions Limited.
[l]A squeegee is a scraping implement with a rubber-edged blade on a handle that is used for cleaning.
[m]Unit operations—discrete, individual activities that achieve specific outputs such as cleaning, grading, various types of mechanical separation (filtration, sifting, centrifugation, distillation), freezing, refrigeration, cooking, etc.

[n]Our experience is that this remains a common practice in multiple plants up to the current time and is a practice that FSQS professional should look out for during assessments as it is often the source of unexplained contamination.

[o]Again, for producers in developing (and indeed even in developed) countries, documentation has remained a weakness and a universal problem for the food industry, hence the current minimum requirements are being insisted on by the GFSI set of standards, the EU requirements and the Canadian Food Inspection Agency (CFIA), Australia New Zealand Food Authority (ANZFA) and the FDA and, particularly FSMA regulations.

[p]A good example of this was the scheduling and attendance of the General Manager and Production Supervisor at the FDA-approved Better Process Control School for people involved in thermal processing that was held in Jamaica in January 1998.

[q]This is the important process of standardization of formulations which many developing country processors manufacturing traditional products struggle with.

[r]These were basic requirements for a scheduled or standard process that could be verified as being effective in delivering commercially sterile product.

[s]At the time of the initial assessment, the staff of the facility were using clean cloths to wipe each bottle of beverage to remove any excess product or water and also thereby dry the bottle. This practice was found to be a potential source of contamination.

[t]A hermetic seal is one which is specifically designed and able to keep product in and prevent the entry into the product of microorganisms from the exterior of the packaging (in this case, the bottle). Hermetic means "airtight."

[u]In the author's laboratory, he, as have others, has had cause to use light and light intensity, the presence and concentration of ethylene and oxygen gases as well as differential incubation temperatures as modes of failure in accelerated shelf life testing regimes.

[v]It has been TSL's experience that the best and easiest time to implement food safety and quality systems is when production at a new facility is being commenced. The firm is starting with no preexiting conditions and can do things the right way from day 1. This is highly recommended.

[w]This is a competitor's product from Barbados. Lovermore's seamoss is made by another microenterprise firm with which TSL worked in the late 1990s but has limited distribution within Barbados while Benjo was being exported to more than 12 countries.

Case study: food safety and quality systems implementation in small beverage operations— Mountain Top Springs Limited

A. Gordon
Technological Solutions Limited, Kingston, Jamaica

Chapter Outline

Food Safety and Quality Systems in Developing Countries. http://dx.doi.org/10.1016/B978-0-12-801226-0.00004-9
Copyright © 2017 Elsevier Inc. All rights reserved.

Introduction

Beverages are an important component of the diets in all continents. The range and variety of beverage offerings have changed over the years because of the advance of beverage technology that has allowed man to produce, preserve, and be able to make available a range of beverages from fruits, dairy products, roots and tubers, and a variety of natural and manufactured sources. Of all of the beverages consumed globally, none is more important than water. This critical input to life as we know it has been sourced from rivers, lakes, springs, glaciers, melting snow, and other groundwater sources from time immemorial. Where there is a dearth of fresh water, many countries have found ways to extract water from the sea through a range of desalination technologies. Whatever the source of water, it needs to be safe for drinking and use in the preparation of foods. This means that water typically needs to be treated so that it can meet this basic need. While most developed countries and many developing countries have long ago addressed their populations' basic need for a safe supply of drinking water, there are others that continue to face challenges in this regard. These are coming into focus more as natural disasters and global warming, among other uncontrolled factors, are compromising water availability and safety globally. Events, such as Hurricane Sandy, which left parts of the North Eastern seaboard of the United States without power and safe, potable water for weeks in October 2012; disasters, such as Typhoon Haiyan that devastated the Philippines in November 2013; and the earthquake and tsunami that struck Japan in 2011 are examples of events that had serious consequences for water availability. However, while availability is critical, even more important is that the water is safe to drink, as major foodborne illness outbreaks due to contaminated water have shown that water availability without safety can be devastating (Clark et al., 2003; Hrudey et al., 2003; Ontario Ministry of the Attorney General and O'Connor, 2002). Water is critical is central to life as we know and will continue to be important in the future. This case study will examine how a small beverage company, Mountain Top Springs Limited (MTSL), transformed itself to become a producer of choice for high-quality bottled springwater to several countries in the Caribbean, and how this can serve as a model for other bottled water producers throughout the developing world.

Background to bottled water exports and Mountain Top Springs Limited

Bottled water exports

The trade in water has been one of the fastest growing areas of global food and beverage trade over the last 5 years and was valued at US $3.65 billion in 2014 (ITC Trade-Map, 2015). Of total global exports, developing countries, such as Fiji, Malaysia, and Mexico rank among the top 15 exporters, with Fiji and Malaysia being net exporters. There are many different kinds of drinking water, many of which are being offered in robust containers (typically bottles) for drinking at the convenience of the consumer. The source, constituents, and treatment of the water determine what category it falls

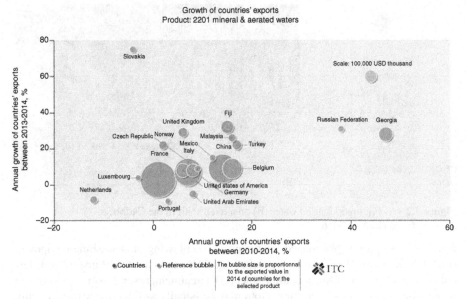

Figure 4.1 Global trade in mineral and aerated waters.
Source: ITC Trademap (from the COMSTAT database).

into: mineral, spring, purified, fortified, carbonated, etc. As the availability and quality of water in a convenient form has become increasingly important, its trade has
increased dramatically, with some categories of bottled water, such as mineral and
carbonated water, growing at a significant pace over the last 5 years (Fig. 4.1). While
traditional exporters of this format, such as, France has seen low-growth rates, relatively new bottled water exporters from Eastern Europe have experienced an explosion
in exports, with Russia enjoying over 32% annual growth and Slovakia 75% over the
last few years. From developing countries, the leader in this category has been Fiji,
which is a significant exporter of its nationally branded, eponymous bottled artesian
water "Fiji," available around the world (Fig 4.2) and which has been growing at a rate
of over 35% per annum over the last 2 years. For all of the producers in these countries
and their product, the safety and perceived quality are paramount to the growth experienced in both their domestic and export markets. This makes the implementation of
sustainably effective food safety and quality management systems critical to bottled
water producers seeking to supply their products to more sophisticated consumers in
their domestic markets or to capitalize on the growing global trade in water.

Production systems

There is a wide range of water-bottling operations in developing countries, ranging
from those bottling high-quality spring or artesian water to numerous beverage manufacturers located across the globe that produce purified water. The complexity of the
processes and, consequently, the type of equipment used varies widely across the industry, depending on the specific circumstances of the firm and the market. Hence, a

Figure 4.2 (A) Fiji brand bottled water (B) on a retail shelf in the Caribbean.
Source: André Gordon, 2016.

firm might employ sophisticated treatment, bottle blowing, and washing equipment (Fig. 4.3) or simply, basic but a very effective combination treatment involving different kinds and levels of filtration, ultraviolet (UV) treatment, reverse osmosis (RO), and manual or semiautomated bottling. Both may be equally as effective and the quality of the finished product will depend more on the effective deployment and use of the systems employed and the vigilance and attentiveness of the quality assurance (QA) team rather than on the equipment.

Water supply and the production of bottled water

The water used for bottled water production can originate from a range of sources. The water is usually derived from a natural well, spring, groundwater source, or, in the case of purified waters, from the municipal water supply. Depending on the analytical profile of the source that will capture its typical microbiological, chemical, physical, and radiological composition, the water will be treated before delivery to the plant for storage, prior to further treatment and use. This treatment may involve the use of a series of filters or, in the case of nonmunicipal water, specific targeted chemical treatment of the source prior to filtration. For pretreated city (municipal) water for making *purified bottled water*, a treatment program from source to filling could look like that described by Hach, a major supplier of analytical equipment for water testing, in its Industry Guide: Bottled Water (Hach, 2015), as shown in Fig. 4.4. However, water from a groundwater source is treated in accordance to the requirement dictated by the analyses mentioned earlier.

Typically, water sourced from a groundwater source (well, spring, artesian, etc.) will be pretreated, if required, and then pumped to raw water–holding tanks. This water is then passed through sand and activated carbon filters (typically in series). The sand filter physically removes larger suspended particles and large-molecule impurities, while the carbon filter removes organic contaminants, any residual chemical from the pretreatment, and unwanted odors and colors from the raw water. This then undergoes further "polishing" through a polishing (sediment) filter[a] (polisher) and is

Figure 4.3 Water treatment and bottling equipment in a plant in the Caribbean.
Source: A. Gordon, 2015.

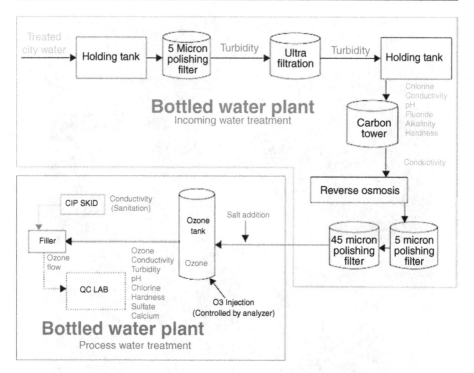

Figure 4.4 Outline of a basic water treatment process for bottled water.
Source: Industry guide: bottled water. Available from: http://www.hach.com/bottledwaterguide.

then transferred into holding tanks, which supplies water for processing in the plant itself or goes directly to the filler holding tank, as captured in the schematic in Fig. 4.5. Alternatively, the water can be passed through the polisher before being fed directly into the process where it undergoes RO treatment prior to being stored. The polisher serves the purpose of removing small-suspended particles that would have evaded the previous filtration steps, as well as any fine rust or other contaminants that may damage the RO membrane filters. The RO unit removes any remaining organic substances, colloidal material, and some bacteria from the water. Further removal or destruction of any remaining bacteria is achieved by the use of further filtration, often through a 5-μm and then a 1-μm, double-stage filter, followed by UV light treatment and ozonation.

The UV light exposure destroys viable microorganisms, a process that is augmented and reinforced by ozonation, which also should prevent the outgrowth of any microorganisms that survive the process. For mineralized water or some bottled waters, selected mineral salts are added to the water, prior to ozonation. This is to achieve the desired level of remineralization and to give a desired balance to the water, as many of the original minerals are sometimes removed during the filtration during the treatment process. The water from the RO and UV treatment is stored in an intermediate tank for ozonated water from which the filler is fed for filling into bottles, which are capped, placed in cartons, shrink wrapped, and sent to storage (Fig. 4.5). It is important to note that while the process

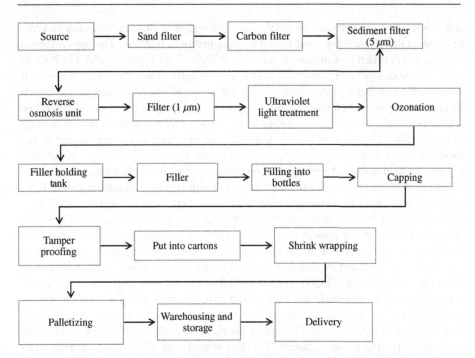

Figure 4.5 Flowchart of a typical process for the production of bottled spring, artesian, well, or other water from a groundwater source.

described earlier (or some variant of it) may be used in many bottled water production facilities, there may be others, such as the producer described in this case study where the purity of the water and its overall a quality is such that minimal processing is required.

Profile of a small beverage producer: Mountain Top Springs Limited

In the Pacific, parts of Central America and the Caribbean, the advent of hurricanes and other weather-induced and natural disasters periodically compromise the availability and quality of the water supply, making it imperative to have an alternative source of safe drinking water, besides the municipal supply. In these scenarios, countries often have to rely on bottled water as their main source of drinking water. In St. Vincent and the Grenadines in the Eastern Caribbean, MTSL plays an important role as a reliable supplier of safe, high-quality bottled water to the people in that country and surrounding islands, particularly in the times of crises. How did this relatively small company in this small developing nation get to a stage where it could supply water that many developed country visitors drink on their sojourn to the Eastern Caribbean? How did it get there? This case study provides the answer.

MTSL, founded in 1994, is a small beverage producer in St. Vincent and the Grenadines that has established itself as a leading producer of high-quality springwater and was the first food production enterprise in the region to attain certification as a recognized quality system for good manufacturing practices (GMPs). This was achieved when it was certified by Canada's Guelph Food Technology Centre (GFTC) in 2002 to its GFTC Auditech GMP Standard. MTSL gets its springwater from a sealed natural spring in the mountains of the Congo Valley, the source of which was capped and sealed by German engineers before the start of production and sale of bottled springwater. The source is located in a remote area with no domestic or farm animal or human activity in the vicinity. The water is gravity-fed down the mountain in an inert piping to the factory where it is treated and bottled. The company, therefore, had the benefit of having an excellent source of high-quality springwater, obtained from a pristine source in an undisturbed part of the world (Fig. 4.6), not dissimilar to the characteristics of Fiji.

Like many small beverage producers in developing countries, MTSL faced multiple challenges in consistently producing high-quality products that could compete with many of the global imports that were available in its domestic market. Among the challenges were the lack of information as to the applicable standards; unavailability of suitably qualified and experienced food scientists, technologists, or other food safety and quality systems (FSQS) specialists; difficulty with sourcing scale-appropriate equipment and, as required, spare parts; and issues with securing a stable supply of production inputs, as needed. Other challenges included finding a suitably trained and educated workforce, the unavailability of the required analytical support, and a weak national supporting infrastructure. MTSL had to deal with all of these. In addition to these, MTSL faced additional challenges of being located in a relatively remote area with limited access to central services and high-transportation costs. Unlike other bottlers, however, that often have issues in finding a secure, safe, accessible supply of high-quality water (from a spring, surface, underground, or municipal water source), MTSL's challenges were outweighed by the undisturbed nature of the source of the water and its freedom from contamination in the immediate vicinity of the recharge area.

After about 6 years of operation, MTSL's management decided that it wanted to significantly expand and upgrade its production to be able to access export markets and better serve the growing hospitality industry in the Organization of Eastern Caribbean States (OECS) subregion. Both of these markets, while impressed with the quality and apparent purity of the water being produced by MTSL, demanded the implementation of systems to assure them that the product could consistently meet globally recognized standards for safety and were manufactured according to GMPs. These demands led to MTSL's pioneering drive to implement a sustainable GMP and hazard analysis critical control points (HACCP)–based food safety system.

Food safety and quality considerations

The World Health Organization (WHO) has identified the basic chemical, physical, radiological, and microbiological characteristics that good quality drinking (potable) water should meet. This is codified in the bottled water standard that it has issued and which it updates periodically (Table 4.1). This standard and its supporting guidelines have a twofold purpose: they seek to limit the presence and possible exposure

Figure 4.6 Mountain Top Springs Limited's (MTSL) factory and its product.
Source: Mountain Top Springs Limited, 2015.

of vulnerable groups to potentially deleterious substances in water, while supporting the inclusion of health-promoting constituents (World Health Organization, 2006). As a guideline value represents the concentration of a constituent that does not result in a significant risk to health over a lifetime of consumption, these values can be regarded as industry benchmarks for safety. Many countries have also issued their own standards for each category of water and major trade organizations, such as the International Bottled Water Association (IBWA), and certification bodies, such as NSF International, have promulgated standards to which they certify bottlers (Table 4.1). The purpose of these standards varies but the standards typically ensure that bottled water is not only safe for consumption, but also generally is healthy for the consumer.

Table 4.1 **Selected parameters from the World Health Organization (WHO), International Bottled Water Association (IBWA), and NSF Standards for bottled water**

Parameters	WHO standards	IBWA standards	NSF international standards[a]
Microbiological standards	mL	mL	mL
HPC	_[b]	-	-
Total coliforms	<1/100	<1/100	<1/100
Escherichia coli	<1/100	<1/100	<1/100
Cryptosporidium	Absent	-	-
Chemical Standards	mg/L	mg/L	mg/L
Acrylamide	0.0005	-	-
Aldrin and dieldrin	0.00003	-	-
Arsenic	0.01	0.01	0.01
Benzene	0.01	0.001	-
Bromate	0.01	0.01	-
Cadmium	0.003	0.005	0.005
Carbon tetrachloride	0.004	0.005	-
Carbofuran	0.007	0.04	-
Chlorine	5.0	1.0	-
Chlorpyrifos	0.03	-	-
Cyanide	0.07	0.1	0.2
DDT	0.001	-	-
1,2-Dichlorobenzene	1.0	0.6	-
1,4-Dioxane	0.05	3×10^{-8}	-
Lead	0.01	0.005	0.005
Mercury	0.006	0.001	0.002
Nickel	0.07	0.1	0.1
Nitrate (as NO_3)	50	10.0	10.0
Nitrite (as NO_2)	3	1.0	1.0
Permethrin	0.3	-	-
Selenium	0.01	-	0.05
Toluene	0.7	1.0	-
Uranium	0.015	-	0.03

DDT, Dichlorodiphenyltrichloroethane; HPC, heterotrophic plate count.
[a]Beverages (general).
[b]Not available.

Potential hazards in bottled water

Several potential hazards have been identified by important standard bearers for water. These include chemical hazards, such as the heavy metals mercury, lead, arsenic, and cadmium; volatile organic compounds and semivolatile organic compounds; synthetic organic compounds; and regulated contaminants, as well as radiological and microbiological hazards (World Health Organization, 2006; IBWA, 2012). The microbial hazards include viruses, such as noroviruses, rotaviruses, and hepatitis A virus; pathogens *E. coli* and *Pseudomonas aeruginosa*; and protozoa *Cryptosporidium* and *Giardia* (Blackburn et al., 2004; Eckmanns et al., 2008; World Health Organization, 2004), the latter being fairly robust and tolerant of disinfectants (World Health Organization, 2006). It must therefore be the goal of all producers of bottled water to eliminate or minimize the presence of these hazards through the application of sound hazard analysis and hazard control practices.

Chemical, physical, and radiological hazards

The major chemical hazards of concern fall within the groups mentioned earlier. Arsenic, uranium, and selenium, which may occur naturally, and lead may give rise to health concerns if they are present in excess (World Health Organization, 2006). Regulated contaminants (e.g., methyl tertiary butyl ether or phenols), volatile organic compounds (e.g., toluene and chloroform), semivolatile organic compounds (e.g., hexachlorobenzene), and synthetic organic compounds (e.g., dioxin and glyphosate) also have to be specifically controlled in bottled water. Further, with regard to chemical hazards, while not typically so considered, the World Health Organization (2006) raised concern about other possible chemical constituents of water. These include fluoride, which occurs naturally but exposure to high levels can lead to mottling of teeth and, in severe cases, crippling skeletal fluorosis. Nitrate and nitrite, when present in water, have been associated with methemoglobinemia, in bottle-fed infants. Their presence may arise from the excessive application of fertilizers in the vicinity of the water source or the leaching of wastewater or other organic wastes into surface- or groundwater. All of these chemicals should be among the parameters for which drinking water is monitored to ensure that they are within acceptable limits in drinking and bottled water (Table 4.1).

Physical hazards found in bottled water typically come from the processing environment or the packaging. They include glass (from glass bottles) pieces of soft or hard plastic, string, flexible packaging, and, occasionally, cardboard.[b] Other physical contaminants, such as pieces of rubber or hard lubricant, are usually the result of lax maintenance practices and postmaintenance cleaning. All physical hazards should be relatively easily controlled by a final filtration step just prior to filling. Their presence in bottled water, therefore, indicates a significant breakdown in practices.

While not common, some bottled water may contain radionuclides that may be naturally present in the source from which they are derived. Any potential health risks associated with these naturally occurring radionuclides should be taken into consideration, although the associated risk is limited because the contribution of drinking water to the total exposure of radionuclides is very small under normal circumstances (World Health Organization, 2006). The recommended approach is to screen drinking water for gross alpha and gross beta radiation activity on an annual basis as required by certain standards (IBWA, 2012; NSF, 2015). This will help in determining whether further investigation of the possible risks is necessary. An important consideration is that the finding of levels of activity above the suggested target values does not necessarily indicate an immediate health risk to consumers (World Health Organization, 2006), as other factors, such as normal background radiation, need to be factored into the assessment.

Microbial hazards

Considering the volume of water (bottled and municipally supplied) consumed globally, the incidences of microbe-mediated foodborne illness due to water are very few. Nevertheless, there have been significant cases reported over the years (Blackburn et al., 2004; Clarke et al., 2003; Eckmanns et al., 2008; World Health Organization, 2006), including several deaths. It is therefore prudent for all manufacturers of bottled water to be aware of the potential microbial hazards and routinely determine whether these may be a cause for concern in the production process or environment. These microorganisms include the bacterial pathogens *E. coli, Campylobacter jejuni, Campylobacter coli, P. aeruginosa,* and *Legionella*; noroviruses, rotaviruses, and hepatitis A virus, and protozoa *Cryptosporidium* and *Giardia* (Blackburn et al., 2004; Eckmanns et al., 2008; World Health Organization, 2004). Other organisms found in bottled drinking water over the years include *Actinomyces*, various species of iron bacteria, and some common mold contaminants that can affect both quality and, potentially, the safety of the beverage.

Microbial illness outbreaks have more commonly been associated with municipal water supplies rather than bottled water, although there have been some notable cases (Blackburn et al., 2004; Eckmanns et al., 2008). *Campylobacter* is a leading cause of foodborne illness in Western countries, with *C. jejuni* and *C. coli* waterborne outbreaks from around the world being typically associated with untreated water. However, the Walkerton (Ontario, Canada) outbreak of 2000 resulted from the entry of *E. coli* O157:H7 and *Campylobacter* spp. from neighboring farms into the town's water supply (Clarke et al., 2003), coupled with inadequate water treatment, and there have been reports of *Campylobacter* contamination from bottled water (Evans et al., 2003).

It is not uncommon for ground- and surface-water sources to have *P. aeruginosa* as a contaminant and *Legionella, Cryptosporidium,* and *Giardia* have been implicated in several outbreaks (Blackburn et al., 2004). In addition, Norwalk, norovirus, and rotavirus (Leclerc et al., 2002), as well as hepatitis A virus are common etiological agents of waterborne illnesses. Care, therefore, needs to be taken to consider and control these potential contaminants in the production of bottled water. Another issue in some bottled waters is the presence of molds, which, though not common, do arise occasionally, presenting as green, white, gray, or black "furry" floating contaminants in bottles. Although usually benign, the presence of mold growth in bottled water (Fig. 4.7A) is unacceptable to consumers. These, as well as the presence of *Actinomyces*, have been found while investigating the occurrence of contaminants in treated bottled spring- and well waters.[c] While not usually a serious cause for concern, some *Actinomyces* are known opportunistic human pathogens and, in rare cases, may cause endocarditis or, more commonly, actinomycosis, which presents as abscesses in the mouth, gastrointestinal tract, or lungs (Adalja and Vergis, 2010; Lam et al., 1993).

A group of waterborne bacteria, which are not known to cause disease, but which can cause undesirable stains, tastes and odors, affect the amount of water the source well will produce, as well as create conditions where other undesirable organisms may grow, such as iron bacteria (Andrews et al., 2013; Minnesota Department of Health, 2015; Rao et al., 2000). Iron bacteria occur naturally in soil, shallow groundwater, and surface waters. They are nuisance bacteria that combine iron (or manganese) and oxygen to form deposits

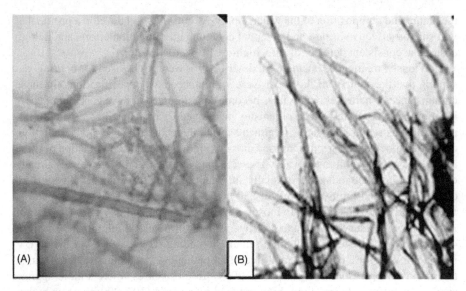

Figure 4.7 Contaminants in bottled water. Scale: X100.
Source: Technological Solutions Limited, 2015.

of "rust," bacterial cells, and a slimy material that helps the bacteria to adhere to well pipes, pumps, and plumbing fixtures (Takeda, 2011; Minnesota Department of Health, 2015) and, in the case of water-bottling facilities, the piping, filler head, filler tips, and other vulnerable parts of the water transmission or handling system. The most troublesome are the filamentous-sheathed members of the order Chlamydobacteriales (Rao et al., 2000). *Leptothrix* is the most widely distributed species of this order, even though *Crenothrix* and *Sphaerotilus* species are also responsible for problems arising from iron bacteria corrosion in cooling water systems (Cullimore and McCann, 1977; Rao et al., 2000). In the work done in assisting a producer to address iron bacteria contamination of production lines, filaments typical of the bacteria *Leptothrix* were observed (Fig. 4.7B). Also found were filaments typical of *Sphaerotilus natans*, though fewer in number. In cases such as these, specific actions need to be taken by the bottler to clean up the system, remove damaged piping, and introduce specific sanitation steps to prevent the reentry or reestablishment of iron bacteria–persistent contamination of production lines and equipment.[d]

Control of potential microbial hazards and contaminants

The mitigation of potential microbial hazards and contaminants in source and bottled waters starts with a proper characterization of the source, transmission, and delivery systems for the water-bottling operations. It should include an examination of the source (if the water is not from the municipal supply) and its surroundings, the transmission system (with a special examination for the signs for the presence of iron bacteria in stainless steel piping), and storage facilities (tanks, etc.). This should be done over a period of at least a year, taking into consideration periods of unusual activity, including weather-mediated events, which, as indicated previously, can change the nature and composition of the water supply. Monitoring will develop a profile that will support effective hazard and potential contaminant mitigation efforts and help to ensure the consistent delivery of safe, high-quality bottled water.

Previously, treatment systems were described for water to be used for bottling, assuming that the sources of the water were within acceptable specifications. In those circumstances where the sources have become contaminated or have gone out of specification because of any of the organisms mentioned earlier, specific, targeted treatments may be required. Specific treatments available include the use of UV radiation, chorine (hypochlorite), ozone, chlorine dioxide, chloramine, or a combination of these to treat the raw or semitreated water. These, along with the filtration processes already described (Figs. 4.4 and 4.5) are usually sufficient to bring the water back within acceptable microbial limits. At levels of 230 mg/L at 0.5°C, pH 7, 99% of all bacteria, viruses, and protozoa *Giardia* are mostly killed by chlorine within 1 min. *Cryptosporidium*, however, survives. Ozone destroys 99% of all potential microbial hazards at 40 mg/L at pH 6.0 and 1°C within a minute, including *Cryptosporidium*, while UV irradiation kills 99% of all potential microbial pathogens, including the protozoa, at 59 mJ/cm^2 (World Health Organization, 2004). In those cases where a more aggressive treatment may be required, high-hypochlorite levels (100–200 ppm) or higher than usual ozone levels (3–4%) may be employed. If treatment is required to eliminate iron bacteria from the bottled water treatment system, a protocol involving the use of sulfuric acid and sodium chloride (NaCl), followed by nitric acid, dibasic ammonium citrate, and ammoniated citric acid is used, at the appropriate pH and temperatures. Testing followed to ensure that there was no acid residue in the system, after which neutralization and routine sanitation were performed.[e]

Quality considerations

Turbidity and chemical balance

Water that is highly turbid, highly colored, or has an objectionable taste or odor may be regarded by consumers as unsafe and may be rejected. It is also accepted that the chemical composition of water affects its flavor, as well as its perceived healthfulness by consumers. This chemical balance is what makes the world's leading bottled water preferred. It is therefore important to be aware of consumer perceptions and to address these quality issues to satisfy consumer preferences.

The improvement opportunity and process and system description

At the beginning of its process of systems implementation, MTSL was using a fairly simple process and equipment to produce bottled springwater (Figs. 4.8 and 4.9). The quality of its product was high due to the attentiveness and the quality focus of the firm rather than the knowledge of FSQS and the formal adoption of any such systems. This, as indicated, would not be atypical of many similar operations, where they exist, across the developing world. It was exposure to information about ways

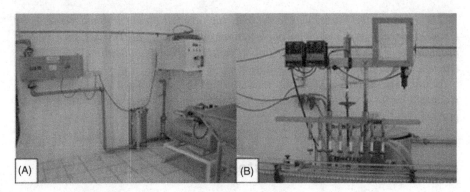

Figure 4.8 Equipment used to produce MTSL bottled springwater in 2001.
Source: A. Gordon, 2001.

Figure 4.9 Overview of the process used to produce MTSL bottled springwater in 2001.
Source: A. Gordon, 2001.

of systematizing quality and ensuring the safety of its product that led to the move of implementing a HACCP-based FSQS, grounded in a certified GMP program along with the attendant prerequisite programs (PRPs).

The specific challenges to be addressed in this case by FSQS implementation were occasional quality issues and lack of the systems required to support the consistent, problem-free production of high-quality, safe bottled springwater.

The specific opportunity was to dramatically expand production and get into se-lected export markets through the implementation of a certifiable HACCP-based GMP program, done in a manner that also significantly improved profitability by improving efficiencies and driving down operating costs.

In this case, a substantial export market for MTSL in Jamaica was identified that would allow for a significant increase in production and the commencement of ex-ports, a first from the firm.[f,g]

Food safety and quality system implementation

The product and original production process

The original process employed by MTS in 2001 is shown in Figs. 4.8 and 4.9 and in-volved the treatment of water gravity-fed to and held in a holding tank, first by passage through a sand filter (5 μm), then a polishing filter (1 μm) prior to further processing. Such was the quality of the water that even in periods following heavy rains, turbidity and the presence of flocculents were never an issue. As such, no prefiltration was re-quired beside the basic treatment described earlier. The filtered water then underwent UV treatment, filtration through another 1-μm filter, followed by passage through a simple RO system (Fig. 4.8A), and then filling into presanitized bottles (Fig. 4.9A) manually loaded onto the filling line (Fig. 4.9B) through a simple pneumatic multipis-ton volumetric filler (Fig. 4.8B). The bottles used (Fig. 4.9C) were blown in the plant from preforms and filled, capped, sealed, and subsequently labeled using a simple manually operated labeler (Fig. 4.9D).

The major quality issues that arose occasionally at the time centered around extra-neous matter (XM) that may be found in bottles and finished product mainly, although there was a single instance of a floating inclusion that appeared to be a mold. As men-tioned earlier, molds are common contaminants of bottled water, if not scrupulously excluded. The production facility has neither had an issue with iron bacteria, nor has there been an instance in which *Actinomyces* or other soilborne bacteria have been found. No food safety issues of any significance have ever arisen for MTSL from its water. *P. aeruginosa* has never been found neither in the source nor at any other point of sampling, or has *E. coli* or any other human pathogen. In the early stages of the HACCP-based GMP systems implementation, in very rare cases, sanitation swabs and isolated internal finished product microbiological quality checks did find the presence of coliforms, the source of which was staff practices and the mitigation of which was a major focus of the systems implementation work.

Characterization of the weaknesses and areas for improvement

Despite its generally good operations, MTSL did not have the systems in place at the outset to satisfy the requirements for certification. From the initial stages of FSQS systems implementation, the plant maintained a clean, sanitary appearance, and good operations. Very little, however, was documented and while there were rudimentary elements of some of the PRPs, these were not necessarily organized in the manner required. No testing of the product, work surfaces, and incoming inputs was being done, largely because the operators were not aware whether these were to be done and, if yes, what was to be done. The location of the plant, as mentioned, meant that there was no easy access to analytical resources and the local analytical support infrastructure was rudimentary, at best. The firm was consistently producing quality products because of its excellent source, the very well-engineered and protected nature of the spring, the pristine environment that has remained free of human or animal contamination, and the scrupulous adherence of the management of the facility to basic good practices. This, of course, was without effective prophylactic systems in place, so there would be unexpected breakdown and occasional failure of controls. These lead to supply interruptions and the need to identify what the source of the particular problem was. Typically, these were plastic or paper in finished product, other XM, or, occasionally, green, white, or other unidentified growth. On rare occasions, coliforms (nonfecal) were found on sanitation swabs.

Development of an approach

To address these issues, MTSL sought technical assistance to help it identify what specifically needed to be done and to develop an implementable program and approach that, if done effectively, would lead to sustainable improvements in its operations. Equally of importance were the perceived business benefits of new markets, which should be a part of the focus of any FSQS implementation process.

The approach taken involved a detailed audit of the systems and operations of the firm. This covered all aspects of the operations from the sourcing, receival, and handling of raw materials and inputs, including the water, the process of bottling springwater, and the supporting infrastructure, including the support (prerequisite) programs. Specifically, this included areas such as:

- training, staff competencies, and competence determination;
- traceability;
- the recall program;
- sampling and testing;
- inventory control;
- pest control;
- standard operating procedures (SOPs);
- sanitation and sanitation SOP (SSOPs);
- labeling and coding;
- storage and transportation;
- preventative maintenance;
- supplier quality assurance (SQA);
- plant layout and design;
- product and personnel flows; and
- the physical premises.

A comprehensive approach was taken that covered all areas of the operations, including offsite warehousing to ensure that any gains made onsite at the plant were not undermined by poor practices elsewhere in the supply chain. The base system chosen for the audit at the time was the GFTC Auditech GMP Certification standard[h] and this was used to underpin the development of an implementation plan that would not only see the firm meeting these requirements, but also address other areas to improve the operations to scale-appropriate industry best practices. The latter was important because, for small firms, it is important that among the available equipment, technology, and systems options, those best suited to current and likely future circumstances and organizational and national culture are chosen. If this is not done, the firm quite often fails to get the benefit of the investment, has on-going system failures, too little or too much capacity, and finds its FSQS, like its other operating systems, not sustainably effective.

The outcome of this very detailed audit was an identification of exactly where the firm was in terms of its compliance with the chosen GFTC[i] standard, it's benchmarking against similar scaled operations regionally and globally, and how it compared with global best practices. The assessment also identified what needed to improve for operations to comply with the requirements of the standard; to develop and implement sustainable, self-reinforcing systems, including a validated HACCP program; and what was required to significantly improve exiting efficiencies, while supporting expanded production for exports. The specific output, in addition to the audit report, was a detailed implementation plan that covered all aspects of the FSQS requirements, as well as the operational and, critically, human resource capability upgrades that would support expansion. The plan covered:

- characterization of the source;
- process characterization;
- plant layout and design (including plant schematics);
- product, personnel, and material flows;
- the implementation of PRPs (described in detail further);
- the development and implementation of a validated HACCP plan for all sizes of bottled springwater; and
- other aspects of FSQS implementation including:
  - the development of a SQA program for all suppliers;
  - control of raw materials: screening of bottles, water, treatment optimization, upgrade of packaging, etc.;
  - control of sanitation;
  - development of a quality assurance program tailored for MTSL's bottled springwater production and distribution operations, including export;
  - development of a detailed sampling and testing program to support the HACCP plan and the quality system; and
  - detailed and specific training of staff to create team members with appropriate knowledge, job-specific skills, a learning culture, and a team-oriented, positive, resilient, and solution-oriented culture.

The plan also covered:

- identification of the root cause of sporadic inclusions in the finished product,
- implementation of solutions for the cause of sporadic inclusions,
- identification of bottlenecks in the production process to facilitate greater efficiencies, and
- reorganization of warehousing and storage to facilitate greater efficiencies and expansion.

The plan was implemented over a period of 12 months with the support of the management and ownership of the firm. Postimplementation, there was a 3-month operation period prior to certification.[j]

General support systems implementation

The process of implementation of the support systems, including the major PRPs followed the plan outlined as a result of the audit that was done. Each program was implemented in a systematic manner to ensure that nothing was overlooked.

Personnel, material, and product flows

The production and process flows were mapped to understand exactly what was going where, the timing of movement of raw materials, inputs, and finished products, as well as what efficiencies and bottlenecks existed. This was coupled with a verification of typical alternate practices occasioned by unexpected occurrences, as well as personnel, product, and process flows throughout the facility, including those under exceptional circumstances.[k] This allowed for a mapping of flows that facilitated identification of any potential areas of problems within the plant and the process itself.

Plant layout, design, and schematics

A detailed assessment of the plant layout and design to understand how the layout of the facility aided or constrained current operations, was undertaken. This was overlaid with the processes, personnel, and materials flows described earlier to give an overview of the operations. The basic layout and flows diagram was similar to the example of a simple layout of a beverage facility (not MTSL) shown in Fig. 4.10. This basic layout was augmented by detailed engineering drawings of the plant, including schematics for all equipment and their locations within the plant; the plumbing, electricals, wastewater plumbing, and flows; and a diagram covering the lighting of the facility. Another schematic outlining the pest control map was also a part of this program. Collectively, these layout, schematics, material, personnel flows, and pest control diagrams documented the facility's operations, providing succinct information to inform decision making.

Characterization of the source

MTSL had chemical, physical, and microbiological nature of the source fully characterized prior to start-up of operations at a laboratory in Germany. Subsequently, this was repeated during implementation and annually after that.

Process characterization

The process used by MTSL for the production of its springwater was mapped in detail, a flowchart of which is shown in Fig. 4.11, and selected aspects in Fig. 4.9. Due to the excellent quality of the springwater and the completely protected nature of the source

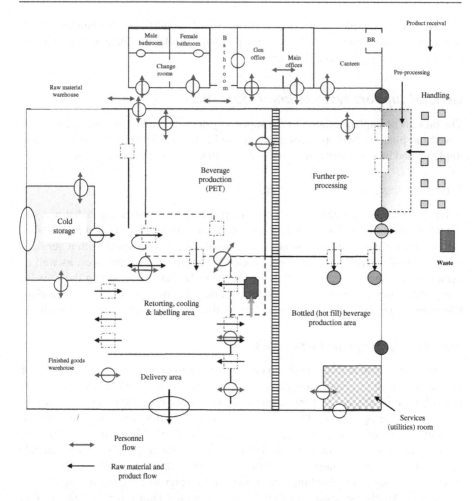

Figure 4.10 Example of draft layout, personnel, and product flow in a small beverage production facility.

and its environment, MTSL did not need to undertake the extensive treatment often done for other bottled waters. Multiple filtration to remove sediments and any flocculents that may be present in the source, followed by carbon filtration, RO treatment, and exposure to UV light was sufficient to ensure consistent safe, high-quality springwater (Fig. 4.11). Ozonation has never been found to be necessary and, therefore, unlike other processes described earlier (Fig. 4.5), it has never been used.

Approach to prerequisite program implementation

Critical to the success of the implementation program was the effective implementation of sustainable PRPs. These laid the basis of the transformation of the operations

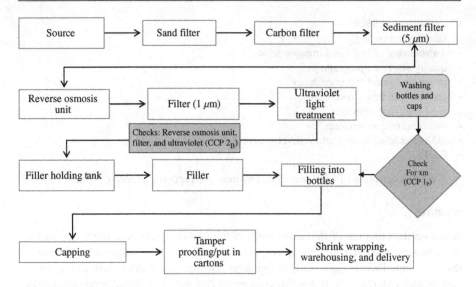

Figure 4.11 Production process for MTSL's bottled springwater. *CCP*, Critical control point; *XM*, extraneous matter.

from one run by the management and the staff on a reactive basis to one based on planned, predicted daily and weekly operations. The objective was not just to implement a FSQS into the operations, but also to transform it to one in which the team had full control of all aspects of the operations and one in which even when the unexpected occurred (e.g., a natural disaster, such as a weather event causing supply interruptions), there would be planned, preagreed approaches to implement solutions: the HACCP-based approach. This HACCP-based approach consisted of identifying the possible areas of problems or failures in each program area and deciding on and documenting mitigating strategies or solutions for these to be implemented should such an eventuality arise. This was a precursor of modern programs that require the development of control points for PRPs, including those, such as the Global Food Safety Initiative (GFSI)–based FSQS programs[1] that require Operating Prerequisite Program (OPRP) control points (Foundation for Food Safety Certification, 2013) or the Preventive Controls rule (Food and Drug Administration, 2015a), also known as Hazard Analysis and Risk–Based Preventive Controls for Human Food (21 CFR 117) under the Food Safety Modernization Act (FSMA). Another related objective was to have the MTSL facility and operations run in a manner equivalent to any other world-class bottled water production operation, utilizing scale-appropriate industry best practices, where they existed, and developing others to meet the specific needs of the business, where required.

The specific PRPs developed and implemented were:

- premises;
- equipment and utilities;
- SOPs;
- pest control;

- sanitation and SSOPs;
- labeling and coding;
- warehousing, storage, and transportation;
- quality assurance program;
- inventory control and traceability;
- product withdrawal and recall;
- personnel/staff training;
- preventative maintenance;
- SQA (including specifications development and monitoring); and
- GMPs.

A brief description of what was done and how each specific program worked follows.

Premises

Based on the outcome of the audit, areas of the premises (external to the production area) that required special care were put on a program of cleaning and maintenance that augmented what was already being done. This included areas, such as the perimeter, drains, external storerooms for equipment (for maintenance of premises, repairs of utilities, etc.), the canteen, and staff facilities. This program became a part of the overall cleaning and maintenance programs of the operations.

Personnel, human resource development, and training

A critical aspect of the operations that was flagged during the audit was the need to significantly upgrade the knowledge of and upskill the workers in the plant, as well as to ensure that the management of the facility had the requisite knowledge to run the operations in the manner required. In essence, this meant that an intensive training and retraining program was required, along with approaches to determine the effectiveness of the training and its impact on both personnel competencies and practices.[m] This aspect of the implementation program cannot be overemphasized, as it has been shown repeatedly that changing the culture of the management and staff of any facility in which a FSQS is to be effectively implemented is critical to long-term success and sustainability (Gordon, 2015; Yiannas, 2008). This required a series of specifically tailored programs covering a range of areas that were delivered in settings appropriate to the circumstances and in which the participants would feel comfortable. Taking factory staff outside of their environment for training can be beneficial, if done at the appropriate time in the program. *If, however, this is done when they are undergoing significant changes to programs and practices, which they do not understand and are uncomfortable with, moving them to an unfamiliar environment where they have to change the way they dress and behave (e.g., a hotel or external training center) could negate the impact of the training entirely.* In such circumstances, doing the training in the production environment (Fig. 4.12A) is preferred.

The training program formed part of an overall human resource development (HRD) program for MTSL that identified aspects of the operations that required specific skills sets and competencies, as well as changes in behavior. If the skills and competence

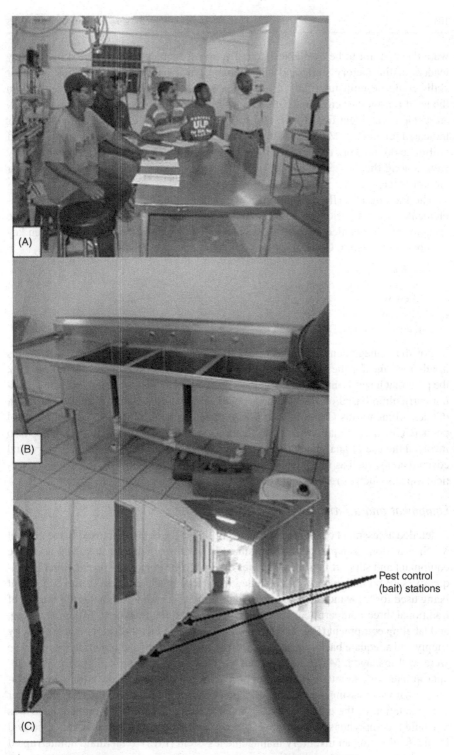

Pest control
(bait) stations

Figure 4.12 Implementation of prerequisite programs (PRPs) at MTSL.
Source: Gordon, 2001/2002.

were deemed not to be available internally, an assessment of the "soft skills" of key leaders of the factory staff[m] and management was performed to determine whether skills could be built internally. In those cases where this was not possible (e.g., in the need for another engineer with responsibility for quality assurance), an appropriate person was identified, hired, and trained to meet the functionality required. As indicated previously, MTSL was located in a remote area, so finding appropriate staff willing to do the work required and also relocate or travel long distances to work daily were among the challenges shared with developing country firms located in similar circumstances.

The training and HRD program for the management and staff of MTSL was implemented over a period of 10 months. Over this period, first basic and then more specialized and detailed knowledge was imparted to the staff in general and then key members of the team. General training (all team members) included:

- basic microbiology,
- basic food and water chemistry,
- food safety,
- GMPs, and
- introduction to HACCP.

For the management and supervisory team, the program went into greater details in all areas and for the HACCP team,[o] detailed understanding of the overall process, the potential hazards, and their mitigation was required. This included a tailored training curriculum based on the standard Codex Alimentarius HACCP training program (Codex Alimentarius Commission, 1985), including hazard analysis, critical control point (CCP) determination, and verification and validation of CCPs. Training also involved the use of analytical techniques and quality assurance, where necessary, and covered areas, such as inventory control, warehousing, and storage, inclusive of practical aspects, such as receival and delivery management.

Equipment and utilities

A detailed assessment of the suitability and utility of the equipment used in the plant at MTSL was done as a part of the assessment to determine whether there was adequate equipment and support facilities (utilities) to enable the consistent production of high-quality, safe bottled water. This included the scale appropriateness of the equipment being used for blow molding; the method of sanitation of bottles, including the use of traditional three-compartment sinks (Fig. 4.12B); and the treatment, filling, handling, and labeling equipment (Fig. 4.9D), among others. The factory had its own electricity supply and adequate backup power, as well as fuel-handling and other facilities. These were well managed. MTSL's operations were found to be well equipped with scale-appropriate equipment that could support growth in production volumes in the future. What was missing was a scheduled preventative maintenance program and better documentation of the existing routine maintenance program. Also implemented was a detailed, comprehensive listing of all equipment, parts, and supplies in, or needed by, the facility and an inventory management system (IMS) to facilitate uninterrupted production operations.

Pest control

Besides routine external spraying, whenever there was thought to be a risk of a pest problem developing, there was no documented or routine pest control program at MTSL at the time of the initial audit. This situation is not uncommon in developing country (and, indeed, in developed country) environments where pest control is not mandated by law or where the availability of suitable providers is nonexistent. MTSL therefore had neither the knowledge of pest control outside of the use of insectocutors to control flies, nor was there a ready source of information to which it could turn. The outcome of the assessment and the subsequent training program introduced MTSL to the requirements for pest control and a range of options for compliance, after which the firm developed and instituted an excellent pest control program that became a model for the industry in the Caribbean region. The program involved protection of the entire perimeter of the building from rodents with the use of bait stations (Fig. 4.12C, arrows) placed no more than 50-ft. apart, as well as internal glue board pest control stations and Vector fly traps. This was supported by a comprehensive set of documentation, covering approved pesticides, the competence of the pest control operator, application sites, and application methods, among other information as outlined in Chapter 1. These were monitored on a daily, weekly, and monthly basis, as relevant. This program was among the first implemented and provided a template for other subsequent PRPs.

Good manufacturing practices

GMPs were among the more important of the programs implemented early in the process. These involved changing the guidelines and acceptable standards for staff attire and protective clothing (Fig. 4.13), the wearing of jewelry, staff practices, signage, handwashing, personnel and product flow, handling and storage of raw materials, basic cleaning, and housekeeping. Procedures were instituted governing personnel matters, staff illness and injury, the management of the premises, waste management, warehousing and storage, the management of glass and brittle plastic, chemicals management, postcleaning and sanitation, and prestart-up inspections, among others. All of this was supported by the appropriate documentation.

Standard operating procedures

An important aspect of the documentation introduced at MTSL was the standardization of all practices, programs, and documentation, including changes to anything that were to be done (e.g., changing of suppliers, approval of new chemicals for use, etc.). This was accomplished through the development and implementation of SOPs for all aspects of the operations. Agreed, standard formats were used for the development of all documentation and SOPs included a description of the program; how things were to be done, by when, by whom, and how this was to be recorded; on which documents and records; how this was to be verified; and how and when the program was to be reviewed. Where necessary, process flowcharts were included as part of the SOPs, if these simplified the procedure.

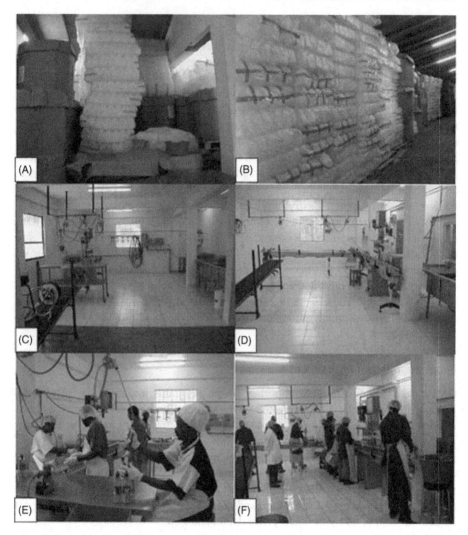

Figure 4.13 Impact of FSQS systems implementation at MTSL.
Source: Gordon, 2001/2002.

Sanitation and sanitation SOPs

An effective sanitation program was developed and validated to ensure that the general environments of the plant and facilities were kept in excellent condition. This included overall master sanitation schedules and other documentation as previously discussed (Chapter 1). In keeping with the development specific procedures to govern cleaning and sanitation, SSOPs were developed and implemented to guide current and future team members in the appropriate and effective ways to undertake any task within the program and how the overall program was to be operated and maintained.

Labeling and coding

An important part of the restructuring of MTSL operations to facilitate the improvement to operate by industry best practices involved instituting detailed procedures for the labeling and coding of all raw materials and inputs in the production process, all in-process inventory and materials, and all finished goods. These included labeling of the areas in which the raw materials and finished products were to be stored and SOPs for how labeling and coding of each item was to be done. The labeling and coding SOPs facilitated the development and implementation of world-class warehousing and storage systems, inventory control, product and input traceability, and recall programs.

Inventory control, traceability, product withdrawal, and recall

The implementation of a new integrated, detailed labeling and coding system dramatically improved inventory control, warehousing (Fig. 4.13A–B), and traceability and facilitated the development and implementation of an efficient system of product withdrawal and recall. This new system easily allowed MTSL to locate anywhere within its operation as little as one case of any input, in-process material, or finished item. It also allowed MTSL to trace any number of cases from any stock-keeping unit to its location and, if necessary, withdraw or recall it, within 2 h, wherever the product was (local or exported). This was in keeping with the most stringent requirements of buyers and global standards and import regulations.

Warehousing, storage, and transportation

In tandem with the improved inventory control and labeling of all items throughout the operations, MTSL was also able to significantly improve its warehouse and storage practices. Limited by a relatively small warehouse, MTSL, through systems implementation and improvement in practices, was able to transform its warehousing and storage practices from those depicted in Fig. 4.13A to the outcomes delivered in Fig. 4.13B. This facilitated controlled loading for transportation for delivery to customers in the domestic market, as well as preplanned picking for loading of export containers. Preloading inspection and preapproval of goods for shipping was routinely done, allowing for offsite planning of warehousing, storage, and inventory management such that the order fulfillment process became significantly easier for all those concerned. Documented inspections were also instituted for all incoming deliveries of raw materials, chemicals, and other inputs.

Supplier quality assurance

Improvements in all of the basic programs meant that MTSL was in a position to implement better control over suppliers and had the basis on which it could implement a specification-driven SQA program. This program included directions for suppliers regarding the specific items that were being purchased (based on specifications issued to them and agreed by them), monitoring of incoming supplies (all raw materials, chemicals, maintenance spares, inputs, and equipment), and, where required, visits to and audits of suppliers.

Preventative maintenance

MTSL always had good maintenance systems in place and had managed to run a fairly efficient operation before the implementation of the HACCP-based GMP program. However, despite begin able to keep an effective maintenance program going, with two in-house engineers, the firm lacked a *preventative* maintenance program, one which would facilitate the planned removal of a key piece of equipment from operation to allow servicing. Its maintenance program was also not originally documented to facilitate predictive and preventative maintenance, the full recording of what was done or to be done, the removal and use of parts from maintenance stores, and full reconciliation of all equipment parts throughout the plant. This was addressed by the implementation program. It also addressed the need for an overall schedule and a comprehensive list of all equipment and areas of the facility that would need maintenance, as well as the requisite associated parts required to effect maintenance. Implementation of this program proved invaluable when MTSL sought to upgrade its equipment and facilities as business and exports grew.

Analytical capability development

In support of all of its GMP activities and the HACCP program, MTSL needed to significantly upgrade its analytical capabilities. These included being able to test in-house for total and fecal coliforms, total microbial numbers, the conformance of packaging to specifications, and the effectiveness of sanitation, among other things. MTSL deployed, in 2002, an early ATP-based bioluminescence screening technology to assess the effectiveness of its sanitation and cleaning processes, well ahead of most of the world[p] at the time. It also instituted routine chemical monitoring of its source and treated water, and it was incapable of doing onsite analyses as done elsewhere at competent laboratories. Collectively, these enhanced capabilities facilitated in-house quality and safety assurance, effective monitoring of suppliers, and, coupled with effective data collection and deployment, provided a significant part of the technical support required for business expansion.

Process and systems upgrade at Mountain Top Springs Limited

The upgrade of the process and systems at MTSL over the implementation period was significant and transformed the operation. The objective of not only implementing systems that complied with the requirements of the firm's target markets and the GFTC GMP Auditech requirements, but also to do so in a manner that was transformative of the business and its production capabilities was met. In addition to what has already been indicated (earlier), specific process and systems upgrades were achieved, as described further.

Plant layout and design

The initial set of changes resulted in the transformation of the layout and production operations of the plant from what it was originally (Fig. 4.13C and E) to a more streamlined layout (Fig. 4.13D and F) that facilitated compliance with food safety requirements. These changes also markedly improved efficiencies and made a significant difference to the ability of the operation to increase production. This was supported by changes in bottle production, sanitation, and handling. In fact, a major bottleneck that was affecting the efficiency of the operations was identified as the process of providing a consistent supply of clean, sanitized bottles for filling (Figs. 4.9A and 4.13F). This process was manual and involved washing each individual bottle in water with residual hypochlorite (0.2 ppm) and then feeding them individually onto the line to be filled the same day. The problem identified was dealt with initially by decoupling bottle production and sanitation from filling, so that the latter was fed from an ample supply of presanitized bottles. Subsequently, efficiencies were further improved in 2004–05 by the addition of new equipment to semiautomate the process of filing, capping, and packaging of the bottled water (Fig. 4.14A–D).

Warehousing and storage

The process of warehousing and storage, its impact on inventory control, and a compliant system of traceability and recall, as well as the management of consumer complaints or returns from the trade, if any, was revolutionized as follows:

1. *Reorganization of the labeling area*: A new room was built and the handling and storage of labels changed to ensure positive release of all labels to specific personnel who were accountable for each label.
2. *Labeling and coding*: This system was implemented as described (under PRPs) earlier such that the positive release of labels into the system, along with the reorganization of the storage area meant that all labels could always be accounted for, as well as all other inputs to the process. This facilitated full traceability and the ability to recall product to meet international best practices.
3. *Reorganization of the warehouse and warehousing and storage*: This resulted in the use of internally developed systems, integrated with best practice solutions brought by the technical support team, which collectively transformed the operation of warehousing and storage, traceability, and inventory management at the firm.[q] The MTSL team developed and implemented an in-house IMS based on a Microsoft Excel spreadsheet that allowed for the location of any inventory within the warehouse within a few minutes. It was supported by a slotting system collectively designed by the MTSL and TSL teams (Figs. 4.13B and 4.14E–F).
4. *Traceability and recall*: The positively released, specific storage in preallocated slots of all inputs, finished goods (Fig. 4.14F, arrows), data collection, and management meant that MTSL had the capability to rapidly trace all inputs and finished good to export markets.
5. *Management of chemicals used and stored in the factory*: This was significantly improved and, as for all other items, under the new IMS, all were accounted for, kept in locked storage with limited access, and all chemicals positively issued, as needed.

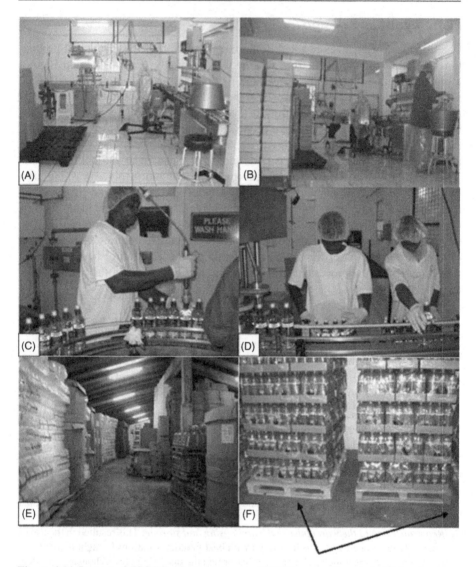

Figure 4.14 Impact of FSQS implementation on MTSL's operations and efficiencies.
Source: Gordon, 2002/2004.

Personnel, practices, and competence determination and building

As previously described, hands-on, plant-training programs, augmented by other training, both on-the-job and in formal classroom settings, were undertaken with the entire production team. All training, competence- and capacity-building activities were followed by a subsequent evaluation to ensure and provide documentary evidence that the staff understood, internalized, and had operationalized the changes in practices required. Practices, such as the placement of buckets containing food contact material directly on the floor and the handling of raw materials and sanitary practices (the placement and routine use of hands-free, handwash stations) were changed. The use of protective clothing on the production floor (Figs. 4.13F and 4.14B–D), as well as the need to change into specific clothing for the production area was also changed. Changes in practices extended to the storage and handling of chemicals (such as sanitizers) in the plant, cleanup of the general operations, and all other aspects of GMPs.

Overall, the changes instituted provided the factory staff with the information, systems, and competence they needed to do their jobs in a standardized, consistently compliant manner. It also allowed the factory staff to essentially run the plant, thus freeing the management to focus on business expansion, a change that paid off handsomely, as will be discussed shortly.

HACCP program implementation

The HACCP program for MTSL was implemented as described in Chapter 1. A hazard analysis of the process, documented by the flowchart (Fig. 4.11), identified two CCPs, which were located after the last treatment step with UV radiation and after sanitation of the bottles and caps, prior to their use. Effective controls were instituted to ensure that all treatment equipment were functioning properly and that there was no XM in the bottles or caps. The program was fully documented, implemented, and scrupulously followed. It was fully supported by the PRPs and other systems that were in place.

Business impact of the implementation: cost/benefit comparison

The question for many food and beverage product manufacturers in developing countries, particularly in circumstances where resources are limited, is often whether allocating these resources to the implementation or upgrading of FSQS is justifiable. This case is therefore a good example where, if properly implemented, the firm can see the payback to the investment exceeding its expectations. In this case MTSL invested approximately US $32,500 in systems implementation and another US $45,000 in equipment and facilities upgrades over the first 2 years (2002–03). The outcome of this investment, besides the attainment of the certification targeted, was major improvements in efficiencies, expansion of capabilities, the access to new markets, improved sales, and profitability.

The improved layout and flows (Fig. 4.14A–B) resulted in significantly enhanced efficiencies, allowing the operation to maximize existing equipment, personnel, and facilities to meet the additional demand. These improvements facilitated targeted equipment upgrades in the blowing, filling, capping (Fig. 4.14C), and packaging areas of the operations (Fig. 4.14D) that had the vicious cycle effect of further increasing throughput and expanding the capacity of the operations. The improvement in the practices in warehousing and storage (Fig. 4.14E), coupled with the use of slotting[r] (arrows, Fig. 4.14F) and an internally developed Microsoft Excel–based IMS allowed for excellent inventory control. Based on positive release with specific individuals, accountable for inventory management per shift, this IMS allowed MTSL to narrow down to the level of an individual case of finished product or input material within a few minutes, anywhere in the warehouse. It also allowed them to be able to trace and locate the product to the point of sale within 2 h, whether released for domestic sale or export.

Critically, the reorganization of the system of inventory management and warehousing and storage allowed MTSL to forego a planned investment in a 2,000-ft.^{2} warehouse expansion in 2002–03 costing approximately US $750,000. It was able to hold the additional expanded production volumes because storage space increased by over 5 times because of the new systems implemented. Additionally, the new IMS meant that MTSL held less inventory per unit produced after the implementation than it did before because of a significantly better understanding and control of raw materials and finished goods inventories. In an increasing number of cases, finished goods went directly into export containers or delivery vehicles, rather than the warehouse, creating room for additional raw material inputs to be kept onsite. Overall, the improvements in efficiency and ability to expand production resulted in driving down the production cost per unit, increasing the profitability of the operations.

The implementation of the HACCP-based food safety system meant that MTSL has had no significant quality assurance issue in the last 12 years and, importantly, no food safety issue. It also allowed the firm to garner contracts to pack for major regional brands, the major one initially being one of the larger distribution companies, based in Jamaica. This opened the door for several other contract packaging opportunities that resulted in MTSL's contract packing business moving from being nonexistent in 2003 to grow significantly by over fivefold by 2008 (Fig. 4.15).

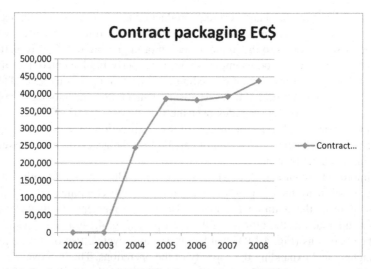

Figure 4.15 Business impact of FSQS implementation on MTSL's contract packing opportunities.

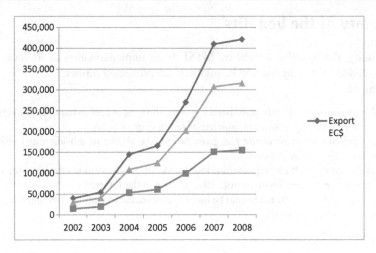

Figure 4.16 Business impact of FSQS Implementation on MTSL's export sales.

MTSL still has significant contract packing business, packaging for a major hotel chain, among others. Importantly, in addition to its contract packaging business,
MTSL was able to dramatically increase exports 10-fold over the 5-year period (2003–
08) postimplementation (Fig. 4.16) to over US $150,000,[s] coming from below 4% of
its total sales of US $574,000 in 2002 to as much as 20% of sales at its peak in 2005.
During this period, the profitability of the operations increased by 163% and, by 2013,
sales had climbed to US $1.37 million. MTSL has shipped its product to five countries
and another eight islands in the Grenadine island chain. Its staff has moved from 5
production and 10 support staff in 2002, when the process started, to 45 by 2013. The
firm has also been able to consistently reinvest in upgrading equipment, production,
and support capabilities such that its asset base has grown by 330% by 2013. MTSL
maintains that all of these achievements have been made possible by the decision it
took to invest in food safety system implementation and the resulting transformation
of its business.

Perhaps the most significant and transformative benefit of the process has been the
revolution in the culture of the staff, and hence the organization, at MTSL such that
the top management no longer directly oversees and runs the operations. It has full
confidence in the staff to discharge its responsibilities, as the process transformed the
core of the staff from being farmers (initially) to modern factory workers, supervisors,
and technical or mechanical specialists. This enduring outcome has laid the basis for
continued future growth of the business. The documentation and record keeping, as
well as the process of having been internationally certified have helped to provide
additional tools to assure the management of adequate maintenance of standards and
practices. This has also been validated by the performance of the business and freedom from food safety and quality issues.

Summary of the benefits

In summary, the benefits derived by MTSL from implementation of its food safety system, divided into tangible and intangibles, are presented further.

Intangible:

- Food safety was not and has never been an issue, allowing access to multiple markets.
- No significant quality assurance issue has arisen in over 10 years
- Robust systems implementation has given MTSL the ability to self-learn and constantly adapt to changing requirements.
- Transformation in culture and human resources meant that the firm has the basis for sustainably expanding production in a competitive manner in the future.
- Additional qualified personnel could be hired to oversee the operations, freeing management to focus on business growth

Tangible:

- Systems implementation has facilitated a significant increase in production volumes and sales in multiple markets.
- MTSL currently ships to 12 neighboring islands, with hundreds of thousands of developed country visitors also being among its market.
- Improved operational efficiencies resulted in significant savings by:
  - allowing MTSL to forego the construction of an additional 2000-ft.^{2} warehouse space, thereby making a significant saving on capital expenditure, which could be directed toward expanding production even further and
  - driving down the per unit production cost.
- Turnover increased significantly by 240% over a 10-year period since implementation.
- Significantly enhanced profitability by the third year after implementation.
- The successes facilitated reinvestment for upgrading equipment, expanding production, and developing and enhancing technical support capabilities.

Conclusions and lessons learned

The process of implementation of a HACCP-based, food safety system for MTSL, driven by GMPs, has shown that even in challenging circumstances in a developing country environment, successful implementation of FSQS is possible and can achieve excellent results. Once the process targets a globally accepted standard and has, at its heart, the change in culture of the firm through an effective, comprehensive program of training and HRD, sustainable transformation can be achieved. Staff at the firm can learn and implement all of the required basic food safety tenets and practices and, if empowered by a management team that trusts and believes in them, attain the required certification and run the operations effectively, freeing management to focus on growing the business. The implementation team should be peopled not just by management and supervisory personnel, but by some of the natural leaders within the operations who must take time to identify, win over, and nurture their colleagues. These natural leaders, more than anything else, will provide the impetus to constantly drive the implementation process, as they have the trust and respect of their peers.

Other important lessons learned is that a proper layout and flow, coupled with an effective, efficient inventory management and warehousing and storage can pay huge dividends: in enhanced flexibility, growth facilitation, and consequently improved profitability. A properly designed and planned implementation program that is target based and driven by important milestones, which are shared with the implementation team, is an effective way to manage project execution. Also, a structured, methodical approach to implementation, which is sensitive to cultural and environmental realities and which seeks creative but globally acceptable solutions to each challenge encountered during implementaion, is the best way to ensure success. When properly implemented, the system can lead to increased market access, significant improvements in efficiencies, enhanced profitability, and sustainable global competitiveness.

Endnotes

[a]Pore sizes here may vary but a commonly used size is 5 μm.
[b]Technological Solutions Limited (TSL) and our associated partner laboratories have encountered several of these potential hazards in bottled water from multiple countries over the last 20 years.
[c]These have been found and positively identified in our laboratories at TSL over the years when working with bottled water from several countries. While *Actinomyces* was positively identified to the generic level, speciation, at the time, was difficult. No specific identification was done for the molds found.
[d]At TSL, we found that having discovered the presence of iron bacteria at one facility, it was best to pull the lines down and undertake a thorough and aggressive cleanup, going all the way back to the well to eliminate or minimize the presence of the organisms in the system. Otherwise, a persistent infection, which could lead to nonpathogenic but esthetically displeasing sporadic contamination of product, could compromise the brand may result.
[e]A protocol based on this sequence of application of agents was successfully employed by TSL to clean up a pervasive, highly resistant systemic contamination of a bottled water line with iron bacteria.
[f]Mountain Top Springs Limited (MTSL) had never exported bottled water before but other products were being shipped to Jamaica from St. Vincent and so there was a direct market access possibility.
[g]TSL was hired to help MTSL meet the requirements of the customer in the importing market for assurance that MTSL could consistently meet their quality and supply requirements. Food safety and quality systems (FSQS) implementation and good manufacturing practices (GMP) certification provided the assurance required to the importing firm.
[h]TSL, the consultant firm chosen, had a relationship with Guelph Food Technology Centre (GFTC), hence the selection of one of GFTC's standards. The specific standard was chosen at the time because it was perceived to be flexible and to allow producers from developing countries to meet global best practice standards. Any other suitable standard could have been chosen [such as those from the American Institute of Bakers International (AIBI), NSF International, the National Food Processors Association (NFPA) (now the Grocery Manufacturers Association; GMA), Silliker Laboratories, Cook and Thurber (now NSF International) or Campden and Chorleywood Food, and Drink Research Association (CCFDR), among others].
[i]GFTC has subsequently been acquired by NSF International and is now NSF–GFTC, operating out of Guelph, Ontario, Canada.

[j]Most certification schemes require *at least* 3 months of operational data to determine compliance.

[k]Understanding what happens under "unusual" or "exceptional" circumstances or unexpected occurrences is important as these are the situations that normally cause quality and food safety problems in a beverage plant. These may include issues which cause a re-routing of routine traffic through the bottle blowing area or warehouse or circumstances such as having to source inputs from an alternate supplier because of unexpected, unpredictable shortfalls in supplies, problems with incoming packaging, equipment breakdown, etc.

[l]The Global Food Safety Initiative (GFSI) is a globally recognized set of benchmarked standards used by the food industry to assure the consistent production of safe, quality foods by their subscribers. All of its standards require, to varying extents, that control and effectiveness of the various prerequisite programs (PRPs) are demonstrated through objective evidence.

[m]The author and his team at TSL have repeatedly found that this aspect of the implementation program is among the most crucial because if the training and human resource development (HRD) program are not effective, implementation is difficult and any gains made are not sustainable. This is even more so if the facilities' management is not committed.

[n]They may not necessarily previously have held any leadership or supervisory positions. TSL has, over the years learned to identify the natural leaders among workers and, through the HRD program, work with the management of the firm to formally endow these team members with leadership skills and give them responsibility. This enhances their status within the organization and their effectiveness as change agents, thus further catalyzing change.

[o]At that time (in 2001/2002), GFSI was not yet a reality and hazard analysis critical control points (HACCP) was just beginning to gain acceptance in the US and globally, so the term "HACCP Team", not "Food Safety Team" or "Quality Systems Team" was chosen.

[p]From the mid 1990s, TSL had been employing the use of traditional swabs for monitoring the effectiveness of cleaning and sanitation in its own facilities and multiple customer locations. As bioluminescence technology became available and was being approved for use in sanitation monitoring in the late 1990s, TSL started using it in its operations. The consulting firm was, therefore, able to guide MTSL in the application of this technology when it needed a technological solution for sanitation monitoring in 2002. TSL also worked with the Bureau of Standards in St. Vincent and the Grenadines to develop and implement routine validation of the effectiveness of the ATP-based system using traditional swabs, in support of MTSL's needs.

[q]This was TSL, the firm providing technical and scientific guidance and consultancy services to the implementation project.

[r]"Slotting" means the storage of specific products and raw material inputs with unique production or trace codes, in specifically determined, preassigned, and prelabelled areas within the warehouse.

[s]The currency legal tender that MTSL uses is the Eastern Caribbean dollar (EC$). The US$:EC$ exchange rate was (and remains) US$1:EC$2.71.

Case study: formula safe foods—sauces

A. Gordon, R. Williams
Technological Solutions Limited, Kingston, Jamaica

Chapter Outline

Food Safety and Quality Systems in Developing Countries. http://dx.doi.org/10.1016/B978-0-12-801226-0.00005-0

Formula safe foods

Introduction

Food processing in the modern era has advanced to the extent where the process of preparation of food and the foods themselves have been scientifically characterized. This has provided certainty as to whether or not foods are safe or can be made safe through processing and has formed the basis of their acceptance by food scientists in general and those in regulatory bodies, as safe. This reduction of what was traditional practice in the developed world to science and the codifying of this science such that it is easily transferable across borders and continents has helped to facilitate the acceptance of unfamiliar foods as safe in many export markets, once the importer or exporter can provide scientific evidence to support the safety of their product. Among the areas of food science that are commonly applied in determining safety, and hence acceptability, are food chemistry, food analysis for contaminants, and food microbiology. Expertise in other disciplines such as thermal (and nonthermal) processing, food engineering, and ingredient technology are also often applied to advise decision making.

In developing countries, these disciplines and other areas of foods science, where available, are also used by many processors to characterize their products and ensure compliance with third country requirements to allow them to be sold in those markets. As growth in the trade between developing and more developed countries in food continues to expand and as the cuisine of the latter continues to become more varied, more "exotic" food for which the science surrounding its safety is less certain will provide new and interesting challenges to the "settled" sciences of food processing and food safety. Traditional food science also finds itself challenged to accept other traditional foods as safe, regardless of origin, including shiokara or fugu (from Japan), wichetty grub (Australia), or mopane worms a southern African delicacy when consumed raw or dried (Fig. 5.1). However, other traditional products such as sushi, raw oysters, fermented sausages, beef jerky, and various kinds of raw food have been accepted as

Figure 5.1 Traditional foods considered safe in their native countries: fugu (Japan) and mopane worms (Southern Africa).
Sources: www.sushi-girl.wikia.com and www.moneyweb.co.za.

Figure 5.2 Formula safe traditional products stabilized by hurdle technology: paneer (from India) and lap cheong (from China).
Source: www.ruschikitchen.com and www.phothefunofit.wordpress.com.

safe in the developed world, based on extensive studies. These products also include raw milk cheeses, the confits and rillettes that are popular in France and a condiment, cooking aide, and dipping sauce, soy sauce, which is made in a myriad of ways and has become a part of global diets. The challenge arises the world over where the preparation of many traditional products which continue to be made using traditional methods do not lend themselves so easily to the existing characterizations that have been adopted in more developed country markets.

Among this category of traditional products is a special category in which ingredient technology, thermal and nonthermal processing, and hurdle technology, among others, have combined to deliver safe products which do not appear at first glance to be safe. Hurdle technology is a term first attributed to Leistner, who, with his coworkers, has done significant work over the years at the Federal Centre for Meat Research in Germany (Leistner, 1994, 2000; Leistner and Rödel, 1976). Hurdle technology is a method of ensuring the safety of foods by eliminating or controlling the growth of pathogens, making the food safe for consumption and extending its shelf life through the application of a combination of technologies and approaches. In developing countries, these are often quite simple and involve the use of selected ingredients (ingredient technology), water activity (a_w), redox potential,[a] preservatives, and temperature, where applicable (Leistner, 1994, 2000). This approach has been applied successfully to traditional foods from developing countries for many years, including in the stabilization of products such as concentrated cheese whey in Latin America (Aguilera and Chirife, 1994), paneer, a cottage cheese–based product in India and lap cheong (Fig. 5.2), a raw, unfermented sausage from China (Leistner, 1994). The approach, also known as combined methods technology, has long been widely recognized and accepted in the preparation of traditional foods in Europe and is now accepted in other parts of the developed world, including Australasia and Canada. This is the category of foods known by the US FDA as "Formula Safe Foods."

As is evident, the current trends in global food trade suggest that traditional foods produced by methods which employ alternate, traditional processing technologies will assume growing importance in the future. This chapter therefore explores some better and less well-known examples of formula safe foods and gives practical examples as to how these may be characterized to bring them within the ambit of the science that can be used to demonstrate their safety. These should serve as a guide to the approach that can be taken with other traditional products that fall into a similar category and augments the work already done on hurdle/combined methods technology for traditional products from developing countries but from a more industrial production perspective.

Browning

Introduction

Browning is a colorant that is typically dark brown in appearance and is widely used in British and Caribbean cuisine for baking, adding color to meats, and the making of sauces. Similar in nature to better known browning agents used in cooking, browning replaces the addition of sugar directly into the food and gives much more certainty to the outcome. The traditional product (Fig. 5.3) is very heavily used in the baking of dark cakes and traditional buns, particularly at Christmas and Easter time,

Figure 5.3 Leading brands of browning available in the international market made by the processes described.
Source: André Gordon, 2016

respectively. The product can be found in major retailers in North America and the United Kingdom who have, as their main market, Caribbean nationals or foodservice outlets serving Caribbean and Latin American foods.

The traditional process by which it is made involves a caramelization reaction that is characterized by the degradation (including inversion to glucose and fructose) and dehydration of sucrose (sugar) as a result of the high temperatures used (Kowalski et al., 2013). It involves a range of condensation, isomerization, and fragmentation reactions, with the rearrangement of intermolecular bonds and equilibration between different anomeric forms of selected constituents of the resultant complex mixture of reactants while the process in ongoing (Fig. 5.4). The traditional product has a pH of <4.0 because of the chemistry involved in its preparation. Authors such as Buera et al. (1987), Villamiel et al. (2006), and Kowalski et al. (2013) have studied these reactions, including the central role that hydrolysis plays in caramelization browning.

The reaction is facilitated by an increase in temperature and fall in pH which leads to an increased rate of reaction (Ajandouz et al., 2001). Outputs of this process include

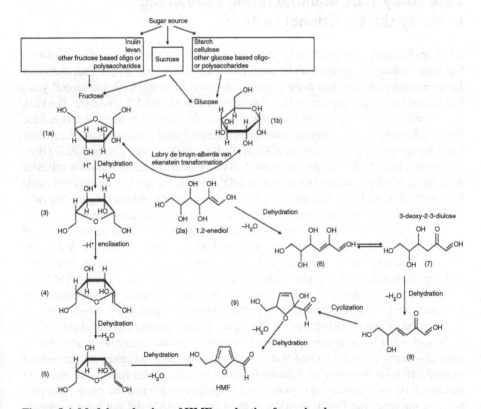

Figure 5.4 Model mechanism of HMF production from simple sugars.
Source: Kowalski, S., Lukasiewicz, M., Duda-Chodak, A., Zięć, G., 2013. 5-Hydroxymethyl-2-furfural (HMF)-heat-induced formation, occurrence in food and biotransformation—a review. Pol. J. Food Nutr. Sci. 63 (4), 207–255.

5-hydroxymethlyl-2-furaldehyde (HMF) (Villamiel et al., 2006) and isomaltol, among other products (Damodaran et al., 2007). The fragmentation (breaking up) of these primary products during the process results in the formation of several acids, as well as other products such as acetoin and diacetyl. Caramel pigments such as those found in browning also contain a variety of hydroxyl groups of varying acidity, all of which together determine what the final acidity and pH of the product is likely to be. Typically, the pH of caramel based mixtures falls below 4.5. Alternatively, browning can be made by a process which involves the use of premade caramel as an ingredient which is blended with other ingredients to make another variant of browning. That process also makes a product that is acidic in nature. This set of case studies examine the approaches taken to characterize two types of browning from two different companies that were unfamiliar to regulators in the United States of America which took prohibition action against them being imported into that country.

Case study 1: UL Manufacturing's Browning made by the traditional method

UL Manufacturing, a manufacturer of browning made by the traditional method who had been making the product under contract for GraceKennedy & Co. Ltd. and other Jamaican exporters who had been exporting the product to the USA for several years had a shipment of their product held at port of entry into Miami, USA in 2002. The FDA Compliance Officer, in assessing the contents of the shipment, indicated that it contained products for which there were no scheduled processes[b] filed with the FDA's Center for Food Safety and Applied Nutrition (CFSAN)'s Low Acid Canned Foods (LACF) division, as required. Without a process established by a process authority[c] for what appeared to be an acidified product and the attendant SID[d] number, the product would continue to be prohibited in the USA. This also affected all other exporters of browning into the market as they all now found their products being held and the same questions being raised.

The challenge faced by the processor was that the product had never been filed because they were unaware that it had to be and what category to have it filed under. This was because the product was atypical of products typically filed as "acidified" and it could not be filed as a low acid food as it clearly would not comply. TSL[e] was contracted to assist and undertake an independent assessment of several production processes from browning commercially available on the market to verify whether or not the product was acidified, whether the product needed to be filed and also to evaluate where it fell in the spectrum of similar products on the Caribbean market which were also exported to the United States. In addition to that, TSL assessed the method of production for browning at different facilities as well as their finished product and undertook its own simulation to verify and properly characterize the production process. At the time, the LACF division of the FDA's CFSAN did not have a specific option on their filing menu for products that were not either LACF or acidified foods (AF), but were made safe through other means. The objective was to provide support documentation to the FDA on the status of the product.

Table 5.1 **Comparison of browning on the Jamaican market in 2002**

Browning	pH	°Brix	Viscosity (cP)	% Acidity	Ingredients (listed)
1	2.59	75	475	1.49	Water, caramel, salt, sodium benzoate, approved spices
2	3.97	71	225	2.28	Caramel, water, approved spices
3	3.90	73	150	2.17	Caramel, water, approved spices
4	3.82	37	6080	1.15	Water, caramel, starch, salt, sodium benzoate
5	2.4	74	600	1.58	Water, caramel, starch, salt, sodium benzoate

Survey of browning on the market

Several samples of browning were taken from the Jamaican marketplace and analyses of their pH, Brix, viscosity, and acidity were conducted (Table 5.1). All of the products, subsequently found to be made by a variety of methods some of which will be discussed in subsequent sections, had finished product pH below 4.0. Of these products, UL Manufacturing's product was among the more acidic at pH 2.59 and had the best natural viscosity (475 cP^f).

Verification of the process by simulation

Samples of browning were collected at different stages of the manufacturing process from UL Manufacturing and assessed at TSL's Testing Laboratory. Ingredients were also sampled from the manufacturer for comparison with an in-house, bench top simulation of the manufacture of the product. The simulation and the comparison with product and inputs from the commercial process was to validate the results obtained from the commercial process and verify that it was possible to get similar results to that obtained commercially without the addition of any acid or other acid ingredient, outside of the caramel that played the role of a "starter/accelerator" of the process. The outcome of the simulation and comparison analyses done are presented in Table 5.2.

The traditional process of browning production developed many years ago in the Caribbean involved heating raw sugar, with constant stirring, until it was molten. The color of the melted sugar was checked visually at regular intervals. When the desired color was attained, water was added and the resulting mixture was cooled and filtered into sanitized high density polyethylene (HDPE) drums. This process was repeated for subsequent batches. Spices and a preservative were added to the drums to complete the product. The drums were then stored until ready for bottling. An outline of this process, which was also used by UL Manufacturing, is given in Fig. 5.5. Different facilities used processes that varied from this traditional process. Some produced

Table 5.2 **pH of commercial and simulated browning and ingredients**

Sample	pH
UL Manufacturing's water used in the product	6.47 @ 28°C
TSL's water used in the product	6.73 @ 26°C
Caramel color (ingredient)	3.20 @ 27°C
Sugar solution (in water)	6.01 @ 24°C
UL Manufacturing's Caramelized Sugar	2.48 @ 26°C
TSL Caramelized Sugar	2.89 @ 29°C
Simulated browning	2.71 @ 26°C

the browning mainly by diluting premade caramel, adding additional ingredients and blending. Some used a pasteurization cook, while others did not. In all cases, observations of the operations at the facilities eliminated the possibility of chance introduction of acid to the process.

As is evident from the findings presented in Table 5.2, browning is a naturally acid food, having an acid pH and a combination of ingredients all of which (except the water) have an initial pH of less than 4.5, even though no acid is added to the product. The United States Food and Drug Administration (FDA) has defined an "acidified food" as a low acid food to which acid(s) or acid food(s) has been added to produce a finished equilibrium pH of 4.6 or below and a water activity greater than 0.85. Based on these findings (Tables 5.1 and 5.2), browning therefore falls within the category of a naturally acid food as defined in the United States Code of Federal Regulations (CFR) 21 CFR 108 and 114 which sets the standard for Acidified Foods (Code of Federal Regulations, 2016a,c). The browning in this case was an acid food to which no acid had been added and therefore could not be classified as an acidified food. This meant that the company should be excluded from the requirement to register its product and file processing information for it for it to be exported to the United States of America.

Analyses of burnt sugar and browning

The results of the analyses of the browning and burnt sugar samples that were treated similarly to how browning is made are presented in Table 5.3. The analyses (pH and acidity) further demonstrate that the browning was naturally acid which arose from the process of burning sugar.

Categorization of products for exemption

In support of the previous conclusions, TSL reviewed and assessed Table 5.4 which outlines the water activity, acidity, and pH combination as considered by the FDA regarding categorization of products in keeping with 21 CFR 113 and 114 (for LACF and AF, respectively). Table 5.4 summarized from the FDA's registration and filing guidance (Food and Drug Administration, 2002, 2012a, 2016a) and for which the

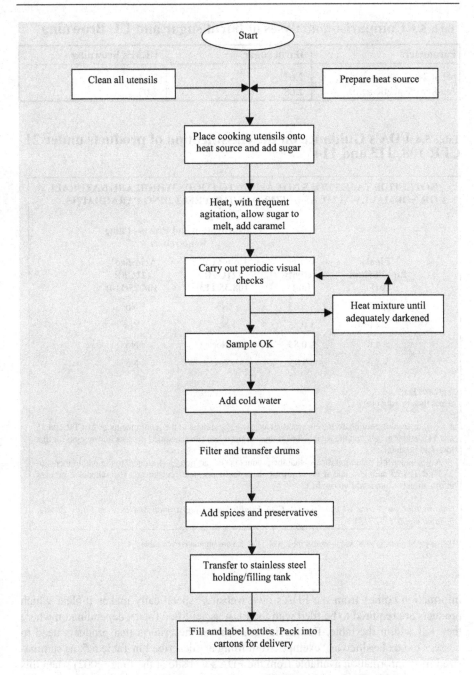

Figure 5.5 General process flow for browning at UL Manufacturing Facility.

Table 5.3 **Comparison analyses of burnt sugar and UL Browning**

Parameters	Burnt sugar	ULM's Browning
pH at 24°C	2.67	2.71
Acidity (% acetic acid)	1.68	2.03

Table 5.4 **FDA's Guidance on the classification of products under 21 CFR 108, 113 and 114[a]**

		Registration and Process Filing Required as:	
Final Equilibrium pH	Water Activity (a_w)	Low Acid* (21CFR 108.35/113)	Acidified** (21CFR 108.25/114)
≤ 4.6	≤ 0.85	No	No
≤ 4.6	> 0.85	No	Yes
> 4.6	≤ 0.85	No	No
> 4.6	>0.85	Yes	No

NOTE: THIS TABLE DOES NOT APPLY TO FOODS WHICH ARE NATURALLY OR NORMALLY ACID, AS DEFINED IN THE PRECEDING PARAGRAPHS.

> greater than
≤ less than or equal to

* A yes under this column defines the product as low-acid, subject to the requirements of 21 CFR 108.35 and 113, and it means that the establishment must register and file scheduled process information for that low-acid product.
** A yes under this column indicates that the product is an acidified food, subject to the requirements of 21 CFR 108.25 and 114, and if so, requiring the establishment to register and file scheduled process information for that acidified product.

(Information was extracted from the US FDA Establishment Registration and Process Filing Booklet updated on 06/17/2013)

[a] The "preceding paragraphs" mentioned at the top of Table 5.3 are summarized in Table 5.4.

information comes from the FDA's own website,[g] specifically makes it clear which products are required to be filed with the FDA as acidified foods, depending on where they fall within the table. Further information on the criteria that products need to possess to be classified as "exempt" from filing are described in Table 5.5, as summarized from information available from the FDA's website at the time (2002) when this case arose.[h] This, along with the information derived from the study undertaken and described in this case study, was sufficient to prove to the importing authorities that the product was safe and fully compliant with their regulations. Although the situation has changed somewhat with an updating of CFSAN's LACF filing system, known

Table 5.5 Categorization of products as "Exempt" from filing under 21 CFR 108, 113, and 114

CRITERIA FOR EXEMPTION FOR FILING PRODUCTS WITH LACF/CFSAN

FOODS NOT COVERED UNDER THE LOW-ACID CANNED FOODS REGULATIONS

(21 CFR 108.35 and 113)

The following foods are not considered low-acid canned foods. Therefore, processors of these foods do not have to register and file processing information for these products:

o   Acid foods (natural or normal pH equal to 4.6 or below)
o   Alcoholic beverages
o   Fermented foods (when the pH of the food is reduced to 4.6 or less by the growth of acid producing microorganisms)
o   Foods processed under the jurisdiction of the meat and poultry inspection program of the United States Department of Agriculture under the Federal Meat Inspection Act and the Poultry Products Inspection Act
o   Foods with water activity (a_w) of 0.85 or below
o   Foods which are not thermally processed
o   Foods which are not packaged in hermetically sealed containers (refer to 21 CFR 113.3(j))
o   Foods stored, distributed, and retailed under refrigerated conditions (product must be conspicuously labeled - e.g., Perishable, keep refrigerated)
o   Tomatoes and tomato products having a finished equilibrium pH less than 4.7

FOODS NOT COVERED UNDER THE ACIDIFIED FOODS REGULATIONS

(21 CFR 108.25 AND 114)

The following foods are not considered acidified foods. Therefore, processors of these foods do not have to register and file processing information for these products:

o   Acid foods (natural or normal pH equal to 4.6 or below)
o   Acid foods (including such foods as standardized and nonstandardized food dressings and condiment sauces) that contain small amounts of low-acid foods and have a resultant finished equilibrium pH that does not significantly differ from that of the predominant acid food. If there is a question about whether a product is covered under the regulations, describe the product, submit a quantitative formula, list pH ranges for each ingredient and submit pH data on finished product from several production lots.
o   Alcoholic beverages
o   Carbonated beverages
o   Fermented foods
o   Foods with water activity (a_w) of 0.85 or below
o   Foods stored, distributed, and retailed under refrigerated conditions
o   Jams, jellies, or preserves covered by 21 CFR 150.

After technical editing is completed on a filing form which has been submitted for a food which falls into one of the above categories, the form is returned to the firm with a letter indicating that process filing for that product or process (such as fermentation) is not required. The manufacturer may be required to provide data to substantiate the exclusion.

as FURLS (FDA Unified Registration and Listing System), Tables 5.4 and 5.5 have proven to be invaluable over the years in dealing with formula safe foods of all types. They are highly recommended for professionals and firms dealing with nontraditional products who are seeking to understand where they fall within the FDA's requirements.

Conclusions

In 2002, when an FDA Compliance Officer in Miami, USA held a shipment of browning manufactured by UL Manufacturing in Jamaica, he asserted that the browning was an acidified foods and therefore needed to be have a process filed with FDA CFSAN's

LACF division. UL's Browning, which has a pH of 2.59–2.71, appeared to fall within the category of an Acidified Low Acid Food, as defined by Codex Standard CAC/RCP 23-1979, Rev 2 (1993) (Codex Alimentarius) and the United States Code of Federal Regulations (CFR) 21CFR 108 and 114 which sets the standard for Acidified Foods. The intervention made on behalf of the firm demonstrated that the product was, in fact, naturally acid and was not artificially acidified. On the basis of the technical information presented the FDA Compliance Officer agreed that the product being manufactured by UL Manufacturing in Jamaica under various brands and being shipped to the United States as browning was a naturally acid food which did not need to be filed with its LACF section, that is, it was exempt. The data and the tables which form an easy guide to the classification of the product and could help to resolve problems at port of entry were recommended to, and consistently did accompany exports of the product made by the firm to the United States. This eliminated the challenge that was being experienced with browning exports to the United States.

Case study 2: Walkerswood Browning and browning made by other processes

Background

The situation that had arisen in 2002 with the browning made by the traditional method with UL Manufacturing's (ULM) product described previously arose again in 2010 with the product from two Jamaican companies and affected all browning exports to the United States. It arose yet again in 2014 with exports from Trinidad and Tobago. The FDA requested evidence of the acidified food (AF) status of the browning made by different processes than the ULM product as well as whether their thermal processes met the filing requirements under 21 CFR 108 and 114 in 2010. A similar request was made of the product from T&T that was detained in 2014. For brevity, all cases will be combined and addressed collectively but differences in the manufacturing processes and products will be highlighted.

Among the product affected in 2010 was browning made by AML Walkerswood (WW), a firm with well-known brands that is among the premier manufacturers and exporters of sauces in the developing country category. WW's product, unlike that of ULM discussed previously, was not made from sugar which had been thermally processed to produce HMF and other nonenzymatic browning (NEB) derivatives (Fig. 5.5) which combined to give the traditional taste, appearance, and properties of browning. The product was made by a much simpler, more direct process in which basic ingredients, including caramel, were mixed and heated together to produce a browning (Fig. 5.6). This process also differed from other processes that also used caramel for making browning (Fig. 5.7) mainly in the ingredients used, the order of addition of the ingredients, and the thermal process applied. Due to the detention of AML Walkerswood Browning at port of entry and the questioning of its safety (i.e., its safety as per the requirements for an approved thermal process filed with the LACF division of CFSAN), a situation that rapidly spread to affect other brands of browning

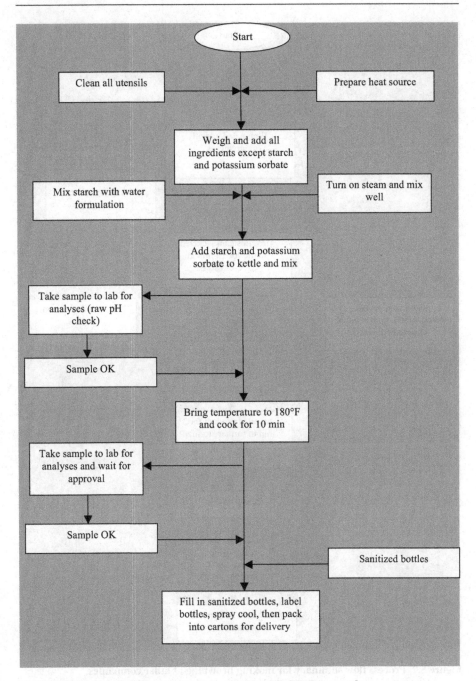

Figure 5.6 Process flow for making browning at AML/Walkerswood.

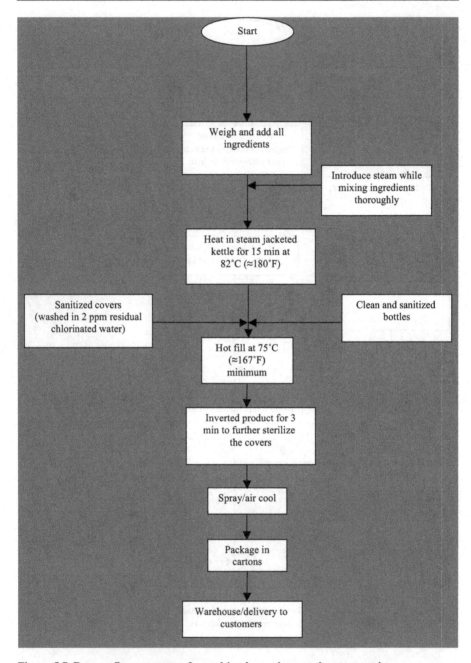

Figure 5.7 Process flow summary for making browning at other companies.

entering the market, WW requested TSL's intervention on their behalf to address the technical requirements of the FDA. Other processors with the same problem also requested assistance to address the matter for them. In all, four processors representing 10 brands were involved in the 2010 situation which became a growing problem, mainly manifested at the Miami port of entry to the United States of America. The 2014 detentions of browning, on the other hand, were from a processor in T&T who also made browning by a process using caramel as a major ingredient (Fig. 5.7) but who was also unfamiliar to the regulators. Again, the port of entry was Miami.

Approach to addressing the problem

The approach taken to address both sets of detention were similar, hence the combining of the cases. In the first instances, discussions were held with the FDA to understand what the issue was, after indicating to them that a similar problem had arisen in the past and had been dealt with by way of evidence as to the classification and safety of the product. It is important to note that although a detailed submission had been made to the FDA on the UL Manufacturing (ULM) case, updated software systems and staff changes meant that the current regulators had no record of the previous filings and had to have the process therefore addressed to their satisfaction. The regulators required:

1. evidence that the product(s) being detained were either essentially similar to what had already been approved (from the previous cases), as well as representation of the documentation on the previous (ULM) case; and
2. alternatively, they wanted to understand the detailed process involved, as well as how the safety of the product had been established and validated.

The documentation required by the FDA meant that an assessment of the production processes for AML Walkerswood's Browning and the other browning products involved had to be undertaken. This was specifically to first document, then validate the effectiveness of the practices employed in manufacturing the browning and delivering a safe product. This required a determination of the appropriate procedures, practices, and process to achieve commercially safe browning.

Browning is a food colorant widely used in Caribbean cuisine and is a naturally acid food, having an acid pH and a combination of ingredients with a starting pH of less than 4.5, even though no acid is added to the product. Browning therefore falls within the category of a naturally acid food as defined in the United States Code of Federal Register (CFR) 21 CFR 108 and 114 which sets the standard for Acidified Foods. In 2002, based on information which we presented to them, the FDA concluded that the product browning coming from Jamaica (including AML Browning) was a naturally acid food which did not need to be filed with its LACF section, that is, it was exempt. Information similar to what was provided to the FDA at that time is discussed in subsequent sections.

The product and the process

The traditional process of browning production involved heating sugar in large pots until it was molten, the color of the melted sugar being checked visually until the desired color was attained, prior to dilution, addition of other ingredients and storage, then bottling. This was the process used by ULM (discussed previously). The process used by AML to make Walkerswood Browning differed from this traditional process and is outlined in Fig. 5.6. Walkerswood Browning is a colorant that is typically dark brown in appearance with a pH of <4.0, making it an acid food. It is made by a process involving mixing ingredients which are heated to facilitate blending, checked for conformance of their pH to specifications, and then cooked at 180°F for 10 min, prior to bottling (Fig. 5.6). The processes used by the other Jamaican manufacturers and for the product from T&T in 2014 are summarized in Fig. 5.7, which is a composite of similar processes.

The classification of AML Browning

An independent assessment of the process of manufacturing WW Browning was undertaken to verify whether or not the product was acidified, whether the product needed to be filed and also to evaluate where it fell in the spectrum of similar products on the market. The ingredients, method of production, as well as the product itself were assessed. The characteristics of the ingredients and product, as well as that from T&T are shown in Table 5.6. Caramel, the main ingredient, is a pigment that contains a variety of hydroxyl groups of varying acidity, all of which together determine the final acidity and pH of the product into which it is introduced, along with any other reactions that may occur. Typically, the pH of the caramel-based mixture prior to final cooking is below 4.0, with the final product pH also being below 4.0 (Table 5.6).

As is evident from the data presented, only the water used in the making of browning by any of the processes applied in the Caribbean is outside of the acid range. The finished products have a pH that is typically below 4.0 and, in the case of WW Browning, substantially so at a pH of 2.7. This product also had a titratable acidity of between 1.2% and 1.38% acetic acid equivalent.[i] The critical factors used in the process for WW and other Caribbean browning products to ensure that the finished product was safe were as follows.

Critical factors

Uncooked blend pH	< 4.5
Finished product pH	< 4.0

Table 5.6 The pH of ingredients and browning from the Caribbean

Sample	pH
Water	6.01 @ 28°°C
Caramel color	4.37 @ 27°C
Uncooked blend of ingredients	3.97 @ 26°C
Walkerswood Browning	2.70 @ 24°C
Browning from T&T	3.97 @ 26°C

Further, for the product held in 2014, the product composition was also provided to the FDA and showed that it was mainly acidified, diluted caramel (49.5%) that was further cooked, resulting in an even lower pH (3.97) than that of caramel itself.

Outcome

These data, in conjunction with the information on the traditional manufacture of browning and how its characteristics compared to those made by other processes were sufficient to establish the safety of the products to the satisfaction of the FDA in both 2010 and 2014. The FDA therefore agreed with the conclusion and submission that *each of the browning in question was a naturally acid food since they each had an uncooked pH of 3.5–3.9 and a finished product pH below 4.6 without further acidification.* They accepted that the products should therefore be *exempt* from filing under 21 CFR 108.25 and 21 CFR 114 and so classified them.

Case study 3: soy sauce as a formula safe food

Introduction

Soy sauce originated in China in the 3rd century but is now ubiquitous in the Far East, being made throughout the region including in Vietnam, Taiwan, Japan, Thailand, Korea, Philippines, Indonesia, and Malaysia, among other countries. Each country has evolved its own traditions and produces different varieties of soy sauce. Although originating in the Far East, the importance of the product in global cuisine has grown such that it is made, used and consumed the world over today, with the Netherlands (at US $147 million) being the major exporter in a US $600+ million industry (Fig. 5.8). Global trade is, however, still dominated by countries from the Asia-Pacific region, as expected, with China, Japan, Singapore, and Hong Kong being among the major exporters (Fig. 5.9). As a result of its ubiquity, soy sauce is found on the shelves of the major retailers in all continents (Fig. 5.10). Also, because of its importance, some of the largest manufacturers and retailers of soy sauce are based in the United States of America and the Netherlands, both of which have become major exporter nations (Figs. 5.8 and 5.9). So important is the product in global cuisine that, despite the many countries and sources from which it can be imported, virtually all continents and regions have their own manufacturers making soy sauce using a variety of approaches.

There are many types of soy sauce, many made in different ways, some from traditional fermentation, others from hydrolyzed soybeans and still others from blending preprocessed ingredients. Due to the wide variations in the way the product can be made, regulators cannot assume that the product should be regarded as safe without understanding the approach used to make it, particularly so when the product is being offered for import. This has been the case with the US Food and Drug Administration (FDA) for soy sauce manufactured in some parts of the world and exported to the United States and forms the basis of this case study of soy sauce handled under the requirements of 21 CFR 108 and 21 CFR 114 of the FDA regulations governing

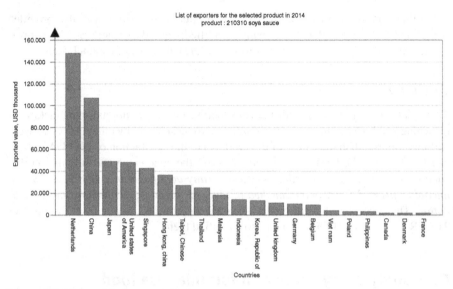

Figure 5.8 Major global exporters of soy sauce.
Source: ITC TradeMap

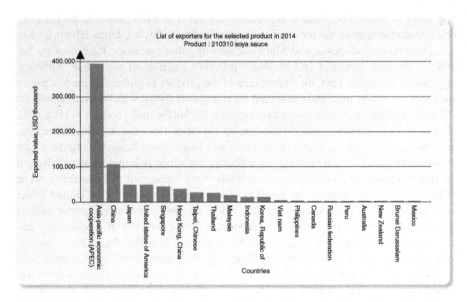

Figure 5.9 Exports of soy sauce from the Asia-Pacific Economic Cooperation Countries.
Source: ITC TradeMap

naturally acid, acidified and formula safe foods. In this section, an examination of how the product is made, the characteristics of various types of soy sauce and how the product is handled by the scientific establishment, as well as regulatory bodies are discussed. Some of the areas of concern for soy sauce manufactured by some of the newer, nontraditional approaches are also discussed.

Figure 5.10 Soy sauce being sold at Walmart, Key Foods and Shoprite in the USA.
Source: André Gordon, 2016

The traditional process for making soy sauce

Various authors have studied the traditional and industrial processes of making soy sauce (Suezawa et al., 2006; Horitsu et al., 1991; Röling et al., 1996) and others have provided insights into the complexity of the processes involved by detailing the biochemistry of the fermentation (Röling et al., 1996; Yong and Wood, 1974; Yokotsuka, 1986). Several have reported on the application of different microorganisms in the industrial fermentation processes involved in the making of soy sauce in various countries (Cho and Lee, 1970; Lee and Cho, 1971; Maheshwari et al., 2010; Yong and Wood, 1974). Government publications in different countries have also sought to promote the benefits of their domestic variety of soy sauce, such as in Taiwan (Chung, 2010), while other studies and publications have documented the nature and benefits of Japanese-style *shoyu*, Korean and Indonesian soy sauce, among others (Kataoka, 2005, Cho and Lee, 1970; Röling et al., 1996). Shurtleff and Aoyagi (2009a,b, 2010) have detailed the history of soy sauce in the Caribbean, South America, and Southeast Asia over the centuries and have documented a comprehensive history of soy sauce from 160 CE to 2012 for the SoyInfo Centre (Shurtleff and Aoyagi, 2012). These sources provide a good insight into the production, diversity, complexity, and importance of this remarkable traditional product that is used in every continent in cooking and flavoring meals.

Traditionally, soy sauce is made by a two-step process which involved an aerobic fermentation stage mediated by molds, followed by a secondary anaerobic fermentation in which yeast and bacteria, including lactobacilli, predominated (Yong and Wood, 1974). In the first stage fermentation, *Aspergillus* spp. (*Aspergillus oryzae*, *Aspergillus sojae*, and *Aspergillus tamari*) are typically used as the main agents of the fermentation and added to the substrate consisting of a mixture of soybeans and wheat. The aerobic fermentation results in a breakdown of protein and carbohydrates, causing a fall in the pH of the mixture to between 4.5 and 4.8 (Röling et al., 1996). During the second and anaerobic phase of the fermentation, yeasts such as *Saccharomyces cerevisiae* and other microorganisms such as *Bacillus subtilis* and other *Bacillus* spp. and lactobacilli, including *Lactobacillus* spp., *Pediococcus* spp., and *Leuconostoc mesenteroides*, among others, further hydrolyze the protein and oligosaccharide components of the soybean/grain mixture to which a brine solution is now added (Lee and Cho, 1971; Maheshwari et al., 2010). *Rodotorula*, *Torolopsis*, and *Saccharomyces* are among the yeasts involved in the fermentation process at this stage as well (Lee and Cho, 1971; Yong and Wood, 1974). This produces the complex flavors, odors, and acidity that typify different kinds of soy sauce. The final fermented mash is then pressed and the liquid (soy sauce) extracted from it, before being heated to inactivate any remaining viable organisms, clarified and bottled (Röling et al., 1996). This is similar to the traditional process that is still used by the major manufacturer of the product in the Caribbean to produce high quality soy sauce that is sold in North America and the United Kingdom. A summary of the process by which traditional soy sauce is made is shown in Fig. 5.11. Good reviews detailing the process and the science behind the production of soy sauce from selected countries are Kataoka (2005), Maheshwari et al. (2010), Röling et al. (1996), and Yokotsuka (1986).

Other alternative processes

The traditional process produces high quality soy sauce through the multistage fermentation involved. However, because it is slow and may take up to 6 or more months for completion (Röling et al., 1996; Yong and Wood, 1974), industrial producers the world over have developed and continue to research ways of making soy sauce in a shorter time (Maheshwari et al., 2010). Some of these newer industrial processes differ from the traditional processes in that they have significantly reduced the time from receipt of raw materials to the delivery of the finished product. Some, such as those still in use in Indonesia, Korea, Japan, the United States, China, Taiwan, and the Vietnam, while having shortened the fermentation time, still rely on the traditional sequence of fermentation using specially selected organisms, more accelerated processing or biotechnology to shorten total production time (Horitsu et al., 1991). Many processors have gone further to eliminate the fermentation phase completely and make a product that is based on blending appropriate ingredients, including a selected acid, prior to heating and hot filling. This has resulted in a wide variety of alternate processes being used, including some that start with a base and add and blend other ingredients to make the final product without any fermentation being involved. One such process is shown in Fig. 5.12.

Although the soy sauce made by these alternate processes is heat processed, some are not regarded as a product that has been made safe because of the pasteurization process, largely because of the formulations used. Such products, as for traditional soy sauce fall within the category of formula safe foods. However, because of the many variations in styles, ingredients and processes used, manufacturers are often required to validate and demonstrate the safety of these products before they are allowed to be imported and sold in some markets, the United States of America being one of these. This case study details the approach for one such product destined for the US market that was required to prove its food safety status.

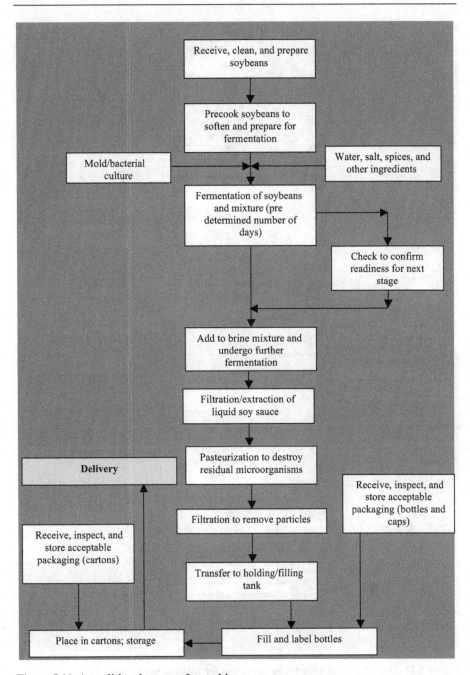

Figure 5.11 A traditional process for making soy sauce.

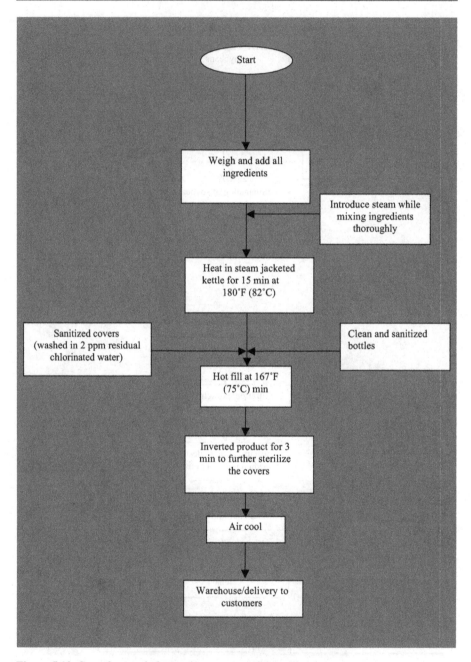

Figure 5.12 One of several alternative processes for making soy sauce.

Background to the case

An exporter of soy sauce to the United States, CBL,[j] found that after several years of shipping without a problem, their range of soy sauces in a container going into that market in mid-2014 were held at the port of entry as the FDA Compliance Officer sought to determine whether the sauces were appropriately filed with the FDA's LACF division of CFSAN. The firm sought technical assistance[k] to provide the appropriate information to meet the FDA's requirements. The soy sauce in question was being made by a nontraditional process similar to that shown in Fig. 5.12, and the challenge was to demonstrate that the product was fully compliant with the requirements under the 21 CFR 108.25 and 114 in the short period of 21 days allowed prior to the container being rejected. The products were essentially made by blending the ingredients, prior to the mixture undergoing a thermal process. The sauce was then hot-filled into 5 oz. bottles prior to cooling, packaging, and warehousing. An assessment of CBL's processing system specifically to determine the appropriateness of the procedures and practices and ability of the process to achieve commercially sterile soy sauce was undertaken, as was an assessment of whether the thermal process applied was required to achieve commercial sterility.

Process for production of CBL's Soy Sauce

The process shown in Fig. 5.12 for the preparation and bottling of soy sauce, as well as the processing practices at CBL, were evaluated for conformance with 21 CFR 108 and 114. It was confirmed that the process, as shown, was the process applied and that specifically:

• The ingredients, consisting of caramel, acetic acid, salt and spices, were added and blended into a mixture and then heated via indirect steam until the temperature of the product reached 180°F. It was then kept at this temperature for 15 min and then strained through a sieve to remove particulates, while still hot.
• The finished product was then bottled at 167°F into sanitized bottles.

It was also confirmed that practices were generally in conformance with the accepted thermal process requirements. The process involved is described as follows:

Recommended process

1. Raw materials are to be washed thoroughly in water containing hypochlorite to deliver a free residual chlorine concentration of 7 ppm chlorine.
2. The appropriate amounts of each ingredient, as per the formulation, should be weighed and added.
3. The ingredients should be thoroughly mixed for 10 min.
4. The product should then be hot filled at 167°F (minimum) into the presanitized glass bottles.
5. The bottles should then be labeled, put in cartons, the latter stored and then distributed, as required.
6. Ensure that the critical factors (see subsequent section) are met.

Measurement on the characteristics of the soy sauce helped to identify the factors that contributed to the achievement of the conditions required for commercial sterility

Table 5.7 **Characteristics for CBL's Soy Sauce initially and after cooking**

Characteristic	Initial mixture (uncooked)	Final
Salt (NaCl) content	—	17.65% (typical minimum)
pH	4.91	4.82
Water activity (a_w)	N/A	0.835 (73.4°F)

in the product. These characteristics included the salt content, pH and water activity, the data for which are shown in Table 5.7.

What was evident from the data and was expected from the formulation used was that the product would be shelf stable (commercially sterile), irrespective of whether it was cooked or not. As this shelf stability would be derived from its low water activity (Table 5.7), the product would fall into the category of formula safe foods.

Recommended thermal process and critical factors

Despite the previous findings, TSL further assessed and documented the thermal process used for preparing CBL's Soy Sauce. This process was found to achieve an adequate level of commercial sterility, regardless of the formula safety of the product. It effectively provides a *5D cook for Escherichia coli*, process adequacy being confirmed by calculations based on the methods of Stumbo (1973)[l] and Breidt et al. (2010).[m] What this meant was that the soy sauce would have been shelf stable with an adequate level of safety on the basis of the thermal process applied, although the product did not require such a process to assure commercial sterility as indicated previously. The critical factors for the product shown in subsequent section therefore exclude the process time and temperature. These must be met to ensure that the appropriate level of commercial sterility in the first instance, based on the formulation used, is achieved. These are presented in Table 5.8.

Status of the product

According to characteristics of the finished product (Tables 5.4 and 5.6), it is not considered to be a low-acid canned food (the final pH is >4.6), nor an acidified food (because a_w is <0.85). Therefore, no registration was required for the product under regulations 21 CFR part 108.25 and 21 CFR part 114. The product, while heat processed, is a food made commercially safe by the nature of its formulation (i.e., a formula safe food) and was filed as such with the FDA, which they accepted. The product has been accepted for import into the United States since then on this basis.

Table 5.8 **Critical factors for CBL's Soy Sauce**

Factor (characteristic)	Level in the finished product
Salt (NaCl) content	17% (minimum)
Water activity (a_w)	0.84 (maximum)

Case study 4: Walkerswood Jerk Seasoning and other cold fill and hold products

Due to the complexity of foods, including particularly traditional foods such as the mango kuchela discussed in Chapter 2 and the Walkerswood Jerk Seasoning being discussed here, the US FDA has had to standardize, as best as possible, its guidance to the industry on acceptable processes for different categories of food. This is typically encapsulated in Industry Guidance documents that can be found on their website (Food and Drug Administration, 2016a). An example of this that is relevant to this case study is the specific industry guidance on Cold Fill and Hold Products (Fig. 5.13), which arose some time after the Jerk Seasoning case had been addressed to the FDA's satisfaction. In this guidance, FDA specifically indicates how acceptable scientific support can be provided for processes that do not involve the heat processing of products that would otherwise be expected to undergo a heating (kill) step (Food and Drug Administration, 2016b). This kind of guidance was not available in 2007 when TSL and Walkerswood (WW) sought to prove to the FDA the safety of Walkerswood's product based on its composition and the attendant water activity that this provided the product with, as well as the GMP used in the product's manufacture. The following case describes the company, its operation, the details of the product, and how it is made, the science behind its safety and stability and the approach taken to convince the FDA of the safety of this product, allowing its entry and sale into the US market.

The Firm: Walkerswood Caribbean Foods Limited

Walkerswood Caribbean Foods Limited is one of the largest producers of sauces in the Caribbean and is based in Walkerswood, St. Ann, Jamaica. Their products are on the shelves of the major multiples and specialty stores throughout the USA, Canada,

Inquiry Codes:

Short Description	Long Description
☑ Challenge study	Please provide scientific support or a challenge study for your cold fill and hold process which establishes how long at a holding temperature it takes for a 5 log reduction of pathogens. The pathogens of concern are E. coli O157:H7, Salmonella, and Listeria monocytogenes.
A Cold Fill and Hold product must be accompanied by a challenge study which shows that relevant acid tolerant pathogens have a 5 log reduction before the product enters distribution. Relevant pathogens include Listeria Monocytogenes, Salmonella, E. coli O157:H7. For Cold Fill Hold products, both a hold time and a hold temperature are critical to the process. The temperature that the product is filled at also needs to be identified. Please refer to a general Cold Fill Hold Study published in the Journal of Food Protection {Vol. 70, No.11, 2007} on Acidified Cucumbers for your reference.	

Figure 5.13 US FDA information for filing a cold fill and hold process.

Figure 5.14 Walkerswood Production Facilities in St. Ann, Jamaica.
Source: Walkerswood Caribbean Foods Limited.

and Europe. The firm operates from excellent facilities[n] (Fig. 5.14) that are well laid
out, well kept and, as a result, along with the production operations and practices, have
allowed the firm to become GMP-certified.[o] Today, the firm operates fully HACCP-
compliant and 21 CFR 117[p]-compliant processes and is constantly upgrading its food
safety and quality systems to meet the new requirements of the global trading environment. Walkerswood Jerk Seasoning (Fig. 5.15A) is made from authentic Jamaican ingredients, including Scotch bonnet peppers (Fig. 5.15B). It is the number one
product in its category of natural, authentic Jamaican jerk seasonings and sauces and
is an award-winning product that is supported by excellent production and quality assurance practices (Fig. 5.16).

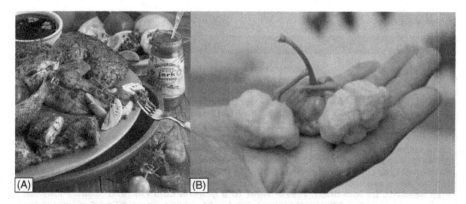

Figure 5.15 (A) Walkerswood Jerk Seasoning with jerk chicken and (B) Jamaican Scotch
bonnet peppers.
Source: Walkerswood Caribbean Foods Limited.

Figure 5.16 Quality assurance activities for jerk seasoning at Walkerswood.
Source: Walkerswood Caribbean Foods Limited.

The Firm: Walkerswood Caribbean Foods Limited

Walkerswood (WW) Jerk Seasoning, as an authentic Jamaican jerk seasoning, is made from a mixture of Jamaican herbs and spices with the primary ingredients being scallion (*Allium wakegi*), onions (*Allium cepa*), thyme (*Thymus vulgaris* L.), Jamaican Scotch bonnet peppers (*Capsicum chinense*), Jamaican pimento (*Pimenta dioica*), and salt, among other ingredients. It is made by a process similar to that outlined in Fig. 5.17 and is further used as a rub or marinade for meats which are to be further cooked prior to consumption. Unlike many other non-Jamaican variants, however, the original, authentic seasoning does not itself undergo a thermal process treatment and therefore it does not have a "kill step" as part of its production process. This was one of the major concerns of the regulators, the FDA, about a product for which they had little information and necessitated the development and sharing with the FDA of scientifically valid information proving the safety of the product to allow it to be sold on the US market. This case study describes the scientific and technical approaches used in 2007 to meet the requirement of proving the safety of the product.

WW Jerk Seasoning is made by a process in which scallion and Scotch bonnet peppers sourced from local farmers are crushed (separately) and stored in a brine solution to be later added to other ingredients in an agitating mixer (Fig. 5.18). The product is finalized by the addition of the requisite ingredients which are mixed together until a homogenous shelf-stable semiliquid paste is produced (Fig. 5.18C). The product is then pumped to a piston filler and cold filled into presanitized glass jars (Fig. 5.18D), capped, labeled, placed in cartons and stored prior to shipment. The process is outlined in Fig. 5.17, as mentioned previously.

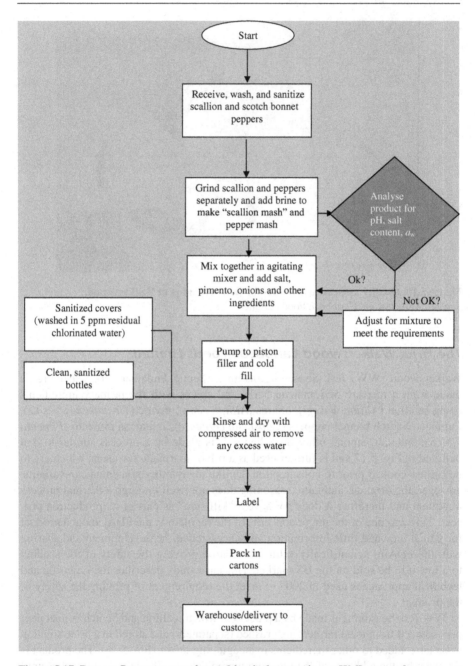

Figure 5.17 Process flow summary for making jerk seasoning at Walkerswood.

Figure 5.18 Process summary for jerk seasoning production at Walkerswood.
Source: André Gordon, 2016

The safety of Walkerswood Jerk Seasoning

As indicated previously, the nature of the product was not one with which the FDA was familiar and, even before issuing the guidance reproduced in Fig. 5.13, they insisted that any exporter of such products into the United States had to provide proof of the products' safety. To meet this requirement, WW approached researchers at TSL and at Iowa State University[q] to undertake analyses and challenge studies on their product to prove its safety. Mendonca (2006) undertook a comprehensive challenge study in which *Salmonella enterica* (5 serotypes), *E. coli* O157:H7 (5 strains), *Listeria monocytogenes* (5 strains), and *Staphylococcus aureus* (5 strains) were used to create separate but mixed-strain inocula which were used to challenge WW Jerk Seasoning. The organisms were inoculated into sterile bottles of the product to give an initial concentration of ~3.0×10^6 CFU^r/g and incubated at 25°C for 7 days. The pH of all 36 samples from both replicates, as well as the water activity of representative samples were assessed. TSL undertook replicate challenge studies with generic *E. coli*, *S. aureus*, and *Salmonella enteritidis* and also undertook three sets of analyses on five batches of product, inclusive of pH, salt (NaCl) content, total bacterial numbers [total aerobic plate count (TAPC)] and water activity.

The batches of WW Jerk Seasoning analyzed by TSL all had microbial counts of <10 CFU/g for the routinely produced product sampled off the line, 10 batches over

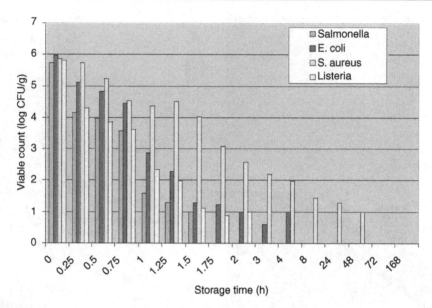

Figure 5.19 Survival of enteric pathogens in Walkerswood Jerk Seasoning at 25°C.
Source: Mendonca, A., 2006. Microbial Challenge Studies to Evaluate the Viability of Four
Human Enteric Pathogens in Walkerswood Jamaican Jerk Seasoning at 25°C. Iowa State
University, October 8, 2006. Submitted to Walkerswood Caribbean Foods Limited, St. Ann,
Jamaica.

a 10-day production period. Further, in years of assessing the product, there had never
been a case of commercially nonsterile product found. No pathogens were found in
any of the routine samples analyzed and for the challenge study, no viable organisms
were recovered after 72 h.[s] Only *S. aureus* was found to survive 24 h and, at 96 h no
viable organisms were found. These concurred with the findings of the Iowa State
University study (Mendonca, 2006) which are shown in Fig. 5.19.

   Mendonca (2006) found that only *E. coli* O157:H7 and *S. aureus* survived in WW
Jerk Seasoning beyond 3 h. Like TSL, he also found that only *S. aureus* survived
beyond 24 h and that beyond 72 h, no survivors of any type were found. In concluding his study, Mendonca (2006) indicated "I can state with a reasonable degree of
scientific certainty that Walkerswood™ jerk seasoning is a hostile environment for
the human enteric pathogens tested in the present study (*Salmonella enterica*, *Escherichia coli* O157:H7, and *Listeria monocytogenes*, and *Staphylococcus aureus*). Also,
based on the use of an initial inoculum level of ~10^6 colony forming units (CFU) per
gram of seasoning, none of the pathogens is likely to survive for more than one week
(168 hours) in this type of product." The typical analytical results for jerk seasonings
(other brands) made by this method,[t] Walkerswood Jerk Seasoning in particular, and
the specific findings of both sets of challenge studies indicate that the product does
not present a risk to human health from the most common pathogens of interest in this
type of product.

Table 5.9 **Typical analytical results for Walkerswood Jerk Seasoning**

Factor (characteristic)	Level in the finished product
Salt (NaCl) content	18% (minimum)
pH	3.16
Titratable acidity (acetic acid)	1.25%
Water activity (a_w)	0.82
Total aerobic plate count	<10 CFU/g
Coliform count	<10 CFU/g
S. aureus	Absent
E. coli	Absent
Salmonella	Absent
Yeast and mold	<10 CFU/g

The typical analytical results of WW Jerk Seasoning resulting from both sets of studies (Mendonca, 2006; Technological Solutions Limited, 2007) and TSL's ongoing analytical results are shown in Table 5.9. These findings over many batches have consistently indicated that the environment in the product was inhospitable for microbial growth, with a pH of 3.16 (on average; always below 3.2), a salt content of 18–20%, added acid to a titratable acidity of 1.25% and an a_w below 0.85. The product, and any such product with a similar composition which has been shown to be inhospitable to pathogens, will therefore always be safe as a result of its formulation.

Outcome and conclusions

The findings of the challenge studies, along with analytical data for 10 batches of product, inclusive of microbiological results, were submitted to the LACF division of the FDA in May 2007. The specialists there reviewed the data and confirmed that they agreed that Walkerswood Jerk Seasoning was commercially sterile and was therefore safe by way of its formulation. This laid the ground work on which similar products could now approach approval by the LACF division and should form a basis on which many traditional products with a similar method of production could approach satisfying the requirements of the FDA for import to the United States of America.

Endnotes

[a]Reduction/oxidation potential—a measure of the availability of oxidizing and reducing agents and their effect on the food.

[b]A "scheduled process" means the process selected by a processor is adequate for use under the conditions of manufacture for a food in achieving and maintaining a food that will not permit the growth of microorganisms having public health significance. It includes control of pH and other critical factors equivalent to the process established by a competent processing authority.

[c]A "process authority" is defined as "the person(s) or organization(s) having expert knowledge of thermal processing requirements for foods in hermetically sealed containers, having access to facilities for making such determinations, and designated by the establishment to perform certain functions."

[d]SID Number—Scheduled Process Identification Number.

[e]Technological Solutions Limited (TSL), the authors' firm.

[f]cP—centipoise—the unit of viscosity.

[g]www.fda.gov.

[h]At the time, there was no such category as "exempt" formally available for filing. This came subsequently. TSL therefore advised exporters to include a copy of the work and these Tables with their shipment to be able to prove to the FDA Compliance Officer at port of entry that their product did not need to be filed and neither would it need to have an SID number.

[i]Titratable acidity is another measure typically used in the industry as an index of the amount of acid available to provide protection to the product against the growth of pathogenic micro-organisms.

[j]CBL is an acronym for the firm whose product is being discussed.

[k]Technological Solutions Limited (TSL) was the firm that provided the technical support.

[l]Stumbo, C.R. 1973. Thermobacteriology in Food Processing. Academic Press, Inc., San Diego California.

[m]Breidt, F., Sandeep, K.P., Arrett, E.M. 2010. Use of linear models for thermal processing of Acidified Foods. Food Prot. Trends 30, 268–272.

[n]The Walkerswood production facilities were built according to a layout and design developed by TSL when the facilities were being upgraded in the late 2000s. They are fully compliant with the requirements in all markets.

[o]Certification is by NSF International.

[p]21 CFR 117 encapsulates the current Good Manufacturing Practices (cGMPs) and Hazard Analysis Risk-based Preventive Controls for Human Foods requirements under the FDA's FSMA Preventive Controls final rule.

[q]Dr. Aubrey Mendonca's laboratory undertook a challenge study covering a wide range of pathogens. TSL undertook a study with *Salmonella*, generic *E. coli*, and *Listeria*. The findings of both studies were similar. The results obtained by Dr. Mendonca whose study was submitted to the FDA in support of Walkerswood's petition for approval for export of the Jerk Seasoning is referenced here.

[r]CFU—colony forming units.

[s]TSL, 2007. Challenge study of Walkerswood™ Jerk Seasoning with *E. coli*, *Salmonella sp.* and *S. aureus*. Internal Technical Paper, TSL's Technology Centre, 237 Old Hope Road, Kingston 6, Jamaica, January 10, 2007.

[t]As found over 20 years of analyses at TSL's laboratory.

Case study: formula safe foods—canned pasteurized processed cheese

A. Gordon

Technological Solutions Limited, Kingston, Jamaica

Chapter Outline

Food Safety and Quality Systems in Developing Countries. http://dx.doi.org/10.1016/B978-0-12-801226-0.00006-2

Introduction

In Chapter 5, a range of sauces and seasoning products that were originally traditional products from developing countries and are becoming a part of global cuisine was discussed. These were shown to be made safe by the use of specific ingredients, formulation, water activity control, or the application of a combination of basic technologies, including thermal processing. The use of combined methods or hurdle technology (Leistner, 2000) has also been applied successfully to the production of traditional dairy foods in some developing countries. The approach, which has long been used in Europe in the production of foods, is now being used in the development and production of dairy products throughout the developing world, including Australasia, Latin American, the Caribbean, the Middle East, and the Far East. These are among the categories of foods classified by the US Food and Drug Administration (FDA) as "formula safe foods." These includes products, such as the concentrated cheese whey found in Latin America (Aguilera and Chirife, 1994) and *paneer*, a cottage cheese–based product that is popular in India and which was shown in Fig. 5.2 in the previous chapter. *Paneer*, which is beginning to gain popularity outside the India subcontinent, is just one of the many traditional dairy products that plays a very important role in enhancing the variety and, more importantly, nutrition of populations of the developing countries. Traditional dairy products that are formula safe or made safe by combined methods technology are the focus of this chapter. A canned processed cheese from Jamaica (Gordon, 2003), which, at least in appearance, is similar to other such products from Saudi Arabia, Austria, Bahrain, and Australia (Fig. 6.1) is an example of such products.

Canned processed cheese is traditionally used in Europe (particularly the Netherlands and Austria) for convenience, with the Austrian dairy firm Woerle and its Happy Cow brand of canned processed cheese (Fig. 6.1) being the European market leader. Canned processed cheese has also been used in other parts of the developed world for emergency preparedness as emergency supplies in the case of storms or other natural disasters, or as a stock item camping supply for outdoor activities, including in the US Army as a standard part of the field rations (Field Ration K, Processed American Cheese). They are particularly useful in situations where refrigerated storage is limited, such as in many areas of Australia where the vast distances and lifestyle of some occupations, such as ranch hands and others, requires them to be away from populated areas, sometimes for many days. Bega is an Australian brand of canned processed cheese (Fig. 6.1) produced by Bega Cheese Limited, Australia. It has a partnership agreement with Fonterra, the firm that is also a 50% owner of the Jamaican firm producing a similar product. Bega brand canned processed cheese has been produced and sold in Australia for more than 50 years.

In the United States, where a canned nonprocessed natural cheese made in Wisconsin, Cougar Gold, is well known, canned processed cheddar cheese still finds a niche. It is imported and sold in outdoor supplies and specialty stores and online retail channels since 2006.[a] Kraft brand canned processed cheese, produced for Kraft, a pioneer in the business, is the brand leader in this category. In developing countries, the production and popularity of canned processed cheese was also occasioned by the

Figure 6.1 Examples of canned pasteurized processed cheese products.
Source: Bega, http://www.begacheese.com.au/; Kraft, shop.khanapakana.com; Happy Cow, http://happycow.at/en/products/processed-cheese/processed-cheese-cans.html; Almari, https:// www.almarai.com/en/brands/almarai/foods/tinned-cheddar-cheese.

need to have safe, nutritious food available in areas with limited access to refrigerated storage, hence the popularity of canned cheese in the Middle East and parts of the Caribbean. Kraft brand canned processed cheese produced in Bahrain, which was originally produced by Kraft Australia, and Almari brand canned processed cheese made by Almari, a Saudi Arabian firm established in 1977 are the market leaders in the Middle East. As access to refrigeration has become more universal, the driver of the demand for canned processed cheese has changed, with it transitioning to become a part of the traditional diet. Now the product is used as a replacement for other cheeses or as a part of traditional delicacies, such as "bun and cheese" (Fig. 6.2B) in the Caribbean.

Although a niche product, the consumption and interest in this dairy product, particularly the unique process by which it is made safe and shelf stable, and the food science behind it is sufficient to merit a detailed examination of the product. This unique dairy product made in Jamaica, a developing country, using a combined methods approach to product stabilization, quality, and safety, and which has been exported to the United Kingdom, the United States, Canada, and other

Figure 6.2 Tastee canned pasteurized processed cheese. (A) Various sizes (B) with the traditional Jamaican "bun", a baked, cake-type product.
Source: www.tastecheese.com. en period?

countries for decades, was prohibited entry into the United States in 2003 by an import alert. The reasons for the imposition of the import alert on Tastee brand canned pasteurized processed cheddar cheese by the US FDA are explored in this first case study in this chapter. As part of this case study, the approach taken to reopen the market for this canned processed cheese product (also called canned processed cheese spread[b]), the science behind the product, and the role it played in resolving the market access interruption are also discussed. The second case study examines the requirements for exporting Tastee canned processed cheese to Canada, a market in which the regulations, as well as the ingredients allowed in such a product differ from the United States. This case study explores the studies required to prove the equivalence of a formulation modified to comply with the Canadian regulatory requirements, including the efficacy of the mix of ingredients used in assuring the safety of the product.

Case study 1: proving compliance of canned pasteurized processed cheese with US thermal processing and safety requirements

Background to Dairy Industries (Jamaica) Limited and its canned processed cheese

In common with many of the products discussed in Chapter 5, the cheese product that is examined in detail in this chapter is a traditional food that forms an important part of the national diet in the countries where it is consumed. As indicated earlier, there are several other canned processed cheese products being sold in the Middle East, parts of Europe, and the Pacific. However, at the time when this case arose, there were no other canned processed cheddar cheese spread products, such as the Tastee-branded product, being imported into North America, the Kraft product only being allowed subsequently in 2006. Made only in Jamaica, this particular type of processed cheese is consumed in Jamaica, throughout the Caribbean, the United States, and Canada, largely by Caribbean nationals living in those countries. It is also exported to other countries around the world. The company that makes the product, Dairy Industries (Jamaica) Limited (DIJL) is a 50:50 joint venture between GraceKennedy, a 94-year-old Jamaican conglomerate and Fonterra (formerly the New Zealand Dairy Board), a more than a century-old global dairy nutrition company owned by 10,500 New Zealand dairy farmers and their families. DIJL produces a range of dairy products, including various cheese, yogurt, and dairy beverage offerings. Tastee brand canned pasteurized processed cheese, its flagship product, has been made by DIJL in Jamaica for over 45 years and exported for over 30 years. It has access to the resources of the Fonterra Research and Development Centre (FRDC) and complements this with the technical support services of Technological Solutions Limited (TSL) for the last 20 years, as it develops new products, improves its world-class food safety and quality systems, and deals with any technical challenges that may arise.

Production process for Tastee cheese

The production process for Tastee brand canned pasteurized processed cheese is shown in Figs. 6.3 and 6.4. The product was made from natural cheddar cheese mainly sourced from Fonterra in New Zealand, which was ground, combined with selected ingredients, pasteurized, and filled and sealed in cans of various sizes (Figs. 6.2–6.4). The process involved the use of salts (NaCl and the emulsifying salts anhydrous disodium and trisodium phosphates) to stabilize the product and facilitate the seamless incorporation of the other ingredients into the cheddar cheese during blending and mixing in the ribbon blenders (Fig. 6.4A) and during thermal processing (Fig. 6.4C). The blended batch was transferred from the blenders to the cooker by the augur cart, infused with steam in the cooker at 92°C for 10 s (0.16 min.), deaerated in the flash vessel (Fig. 6.4C), and then transferred to the filler, all of which is controlled

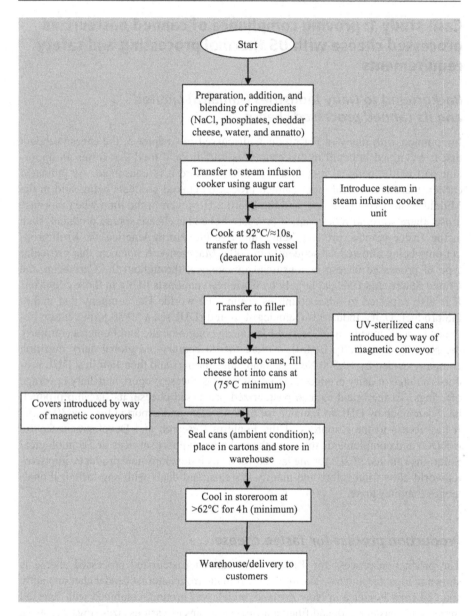

Figure 6.3 Process flow for Tastee canned processed cheddar cheese.

automatically by an operator using a PLC console (Fig. 6.4B). UV-sterilized cans are introduced to the processing lines by conveyors (Figs. 6.3 and 6.4D–E), filled (Fig. 6.4F), and sealed (Fig. 6.4G). The product made has been consistently safe since the inception of its production by this process and had the typical analytical profile shown in Table 6.1.

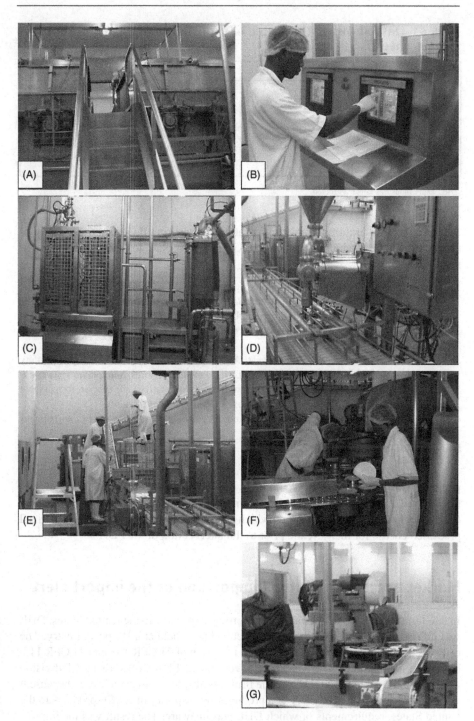

Figure 6.4 The Tastee processed cheddar cheese production process.
Source: André Gordon, 2003.

Table 6.1 Data showing the range of typical analytical values for Tastee cheese

Parameters	DIJL typical Tastee cheese product data
pH	5.4–5.7
Disodium phosphate	2.1%
Sodium chloride	1.7–1.9%
Total salts	3.8–4.0%
Moisture	44–47%
Nisin	0.05%

DJIL, Dairy Industries Jamaica Limited.

There were several important considerations in ensuring the quality and safety of the product, both of which depended on a combination of factors. These can be summarized as follows:

1. *Raw material quality*: The use of good quality raw materials, particularly the cheddar cheese, which should be free of *Clostridium* and *Bacillus* sp., which could cause both economic losses through spoilage and, more importantly, foodborne illnesses.
2. *Effective process control*: Control of the formulation and blending, cooking, filling, seaming, and holding processes.
3. *Lethality and safety margin of the process*: A process with an adequate lethality and safety margin for vegetative pathogens (including *Listeria monocytogenes, Salmonella* spp., *Escherichia coli*, and *Staphylococcus aureus*), as well as any vegetative cells of sporeformers that may be present (e.g., *Clostridium perfringens* or *Clostridium botulinum*).
4. *Control of Clostridium* sp.: The destruction and/or control of the outgrowth of *C. perfringens* and/or *C. botulinum* spores.
5. Adequate and effective record keeping.

Once these factors were effectively managed, sufficient "hurdles" to bacterial pathogen survival and growth would be in place to ensure that the product manufactured by DIJL would be safe, of high quality, and complied with all requirements in the US market. This was the basis on which DIJL manufactured its canned pasteurized processed cheese for many years, prior to the intervention of the FDA.

Market access prohibition: imposition of the import alert

In late June 2003, after two decades of exporting its product to the United States, DIJL was informed that a shipment of its product had been held at a US port of entry. The cause was that the product was found to be in breach of 21 CFR 108 and 21 CFR 113, both of which required the product to be filed with the FDA CFSAN's LACF division as a low-acid canned food (LACF). The firm was also found to be in breach because it was not registered with the FDA as a producer and exporter of LACF products to the United States, requirements of which DIJL was unaware. The result was the imposition of an import alert on DIJL canned processed cheese product until the firm could satisfy the FDA that it met all of the regulatory requirements and that its product was safe. This triggered a process in which DIJL had to use all of the technical resources at its disposal, including FRDC, TSL, and key suppliers (particularly Danisco) to collect the data required to have its product and process fully characterized,[c] and petitioned the FDA for the removal of the import alert.

Dealing with the prohibition of exports

The imposition of the import alert on the imports of canned processed cheese into the United States by DIJL resulted in an immediate cessation of imports into a market, which was critical to the company. At that time (in 2003), unlike today where the FDA has specific guidance on how to address the removal of products from import alerts (Food and Drug Administration, 2016a,b), there was no documented, or at least publicly available, protocols to deal with matters, such as these and so it was left to the ability of firms to engage the regulators successfully to determine whether the specific issue could be resolved in their favor. DIJL sought the assistance of its technical support provider, TSL in resolving the issue. Contact was made with the FDA through TSL whose Process Authority[d] was advised that as a LACF, the firm needed to prove that the product was safe and met all of the requirements for commercially sterility as per 21 CFR 113 (Code of Federal Regulations, 2016b). Due to the complex nature of the product and the fact that it was not sterilized[e] in the traditional sense, DIJL would need to show that its process was delivering a product that was not only free from any potentially hazardous microorganisms, but also that it would remain so during its usable shelf life in the commercial trade. The FDA's LACF Division[f] further advised that on matters to do with the adequacy of complex processes involving dairy products, the FDA would rely on the advice of the Food Research Institute (FRI) and the experts in this area based there, led by Dr. Kathy Glass[g]. DIJL would therefore have to contact FRI's lead specialist and agree on an approach that would suitably address the safety of the product.

Approach to addressing the problem

The outcome of the discussions with the FRI was that DIJL/TSL were required to prepare a technical dossier that would include details of the process, its food safety, and quality system, which managed its production, the details of the thermal process used to make the pasteurized canned processed cheese product, and evidence of microbiological safety of the product. This included the development of a HACCP plan for the product (for implementation at a future date) indicating the organisms for which control would be established, and the nature of these controls, all of which had to be scientifically defensible. Subsequently as the discussions ensued, TSL had to provide verification that the processes established and being practiced by DIJL were sufficiently robust to assure that the thermal process practices prescribed for the product, as well as the food safety hurdles that were used to ensure safety were not likely to be breached. These were based on the hazard analysis that had been done and would for the core of the HACCP plan when implemented in the future. All of this is included in the information presented further that formed the main components of the technical dossier that was submitted to Dr. Glass at FRI and, subsequently, to the FDA as part of DIJL's filing. DIJL was also required to submit samples of several batches of its product to FRI for a range of analyses to be undertaken as a part of its verification of the safety and shelf stability of the product, and to Danisco, its supplier of Nisaplin for nisin determination.

The specific approach to address the issues raised by the FDA involved providing scientific evidence that the thermal process used for the production of Tastee cheese met the requirements of 21 CFR 133.169 for pasteurized processed cheese (Food and Drug Administration, 2015e) and also 21 CFR 113 for LACF (Food and Drug Administration, 2002, 2012a), as the product was not only a pasteurized processed cheese, but was also a LACF. It was also required that the DIJL provide evidence that the food was made safe by way of its formulation, that is, compliance with the Tanaka Principles as outlined in discussions with FRI, as this was necessary to satisfy the FRI/FDA standards. Finally, DIJL also had to provide evidence that it had a system in place to ensure that the parameters outlined by Glass (2003), discussed further, would consistently be met. The details of the information and data presented, as well as aspects of the design of the data gathering that was done are outlined under the appropriate captions further.

Scientific background to Tastee processed cheddar cheese production

Once there was an understanding as to the approach that would allow for the removal of the import alert, DJIL and TSL undertook a carefully crafted program to systematically address all of the issues raised by the FDA through FRI. The exiting process was fully characterized and all relevant procedures that enabled control of the safety and compliance of the product were carefully documented. These included the thermal process, formulation-based food safety controls, and the routine data collection and documentation that ensured that controls were being effectively implemented. The program that was implemented and the information presented to the FDA through FRI is detailed further.

The thermal process

TSL undertook an assessment of the exiting production process for Tastee pasteurized processed cheddar cheese using the protocol outlined further. DataTrace MPII temperature probes (Fig. 6.5A), placed in-line where possible (Fig. 6.5C) and were used to record the thermal processing data. Temperatures of cheese product taken from the augur cart (Fig 6.5B), the cooker (Fig. 6.5C), the flash vessel (deaerator), the filled cans (Fig. 6.5D), and cans during cooling were recorded and are presented in Tables 6.2 (augur cart) and 6.3 (filling), with selected thermal process profiles being shown in Figs. 6.6–6.9. The summary thermal process data is shown in Table 6.4.

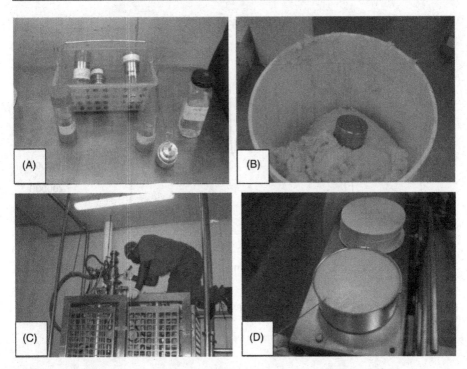

Figure 6.5 Validating the thermal process for Tastee processed cheddar cheese.
Source: André Gordon, 2003.

Table 6.2 **Auger cart temperatures**

Samples	Time of day	Probe	Temperature (°C)
1	11.20	M3T39510	19.4
2	11.39	M3T39504	19.3
3	11.48	M3T39508	19.6
4	11.59	M3T39501	19.6
5	12.08	M3T39506	19.5

Table 6.3 **Filling temperatures**

Samples	Time of day	Temperature (°C)
1	11:32	70.2
2	11:45	72.2
3	11:53	69.2
4	12:16	70.3

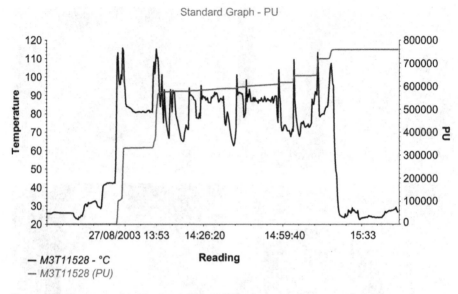

Figure 6.6 Pasteurization during cooking of Tastee canned processed cheese.

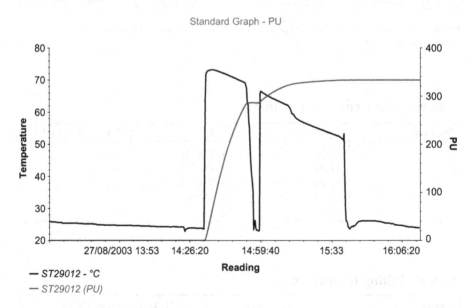

Figure 6.7 Pasteurization during holding of Tastee canned processed cheese.

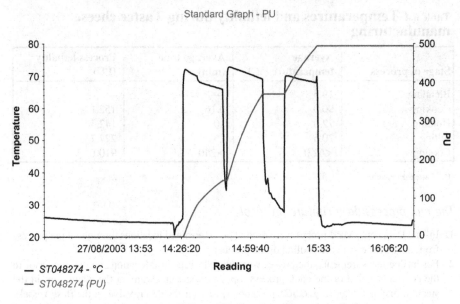

Standard Graph - PU

— ST048274 - °C
— ST048274 (PU)

Figure 6.8 Pasteurization during filling of Tastee canned processed cheese.

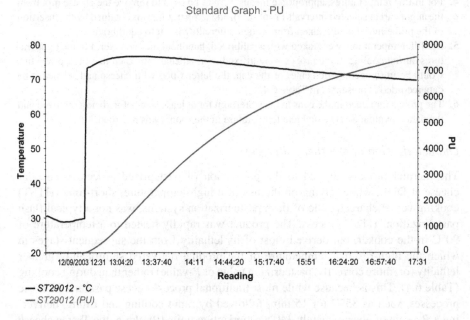

Standard Graph - PU

— ST29012 - °C
— ST29012 (PU)

Figure 6.9 Pasteurization during cooling of Tastee canned processed cheese.

Table 6.4 **Temperatures and lethality during Tastee cheese manufacturing**

Stage of process	Average temperature (°C)	Average time (min)	Process lethality (PU)
Blending	19.5	—	—
Cooking	90.0	0.16	152.1
Holding	72.0	10	142.3
Filling	70.5	16	227.7
Cooling*	<62.0	>240	9100

PU, Pasteurization units.

Thermal process data capture protocol

1. In-line DataTrace MPII probes were used at predetermined locations throughout the process from the augur cart to the cooling of the product.
2. For in-line measurements, the probes were placed in specific locations on the line, prior to the commencement of the cook process. Specific locations were: at the augur cart, in the steam infusion cooker (Fig. 6.5C), postcook (just after the divert valve, at the flash vessel), at filling, and in the cans to monitor the temperature during cooling.
3. The probes remained in place until the end of the production run, at which time they were retrieved, returned, and read.
4. For in-line temperatures, appropriate means were to be used to remove the cheese mix from the augur cart at selected intervals (Table 6.3), and temperature was obtained by the insertion of the probe into this mix taken from trough, after allowing it to equilibrate.
5. For fill, temperatures were taken with a calibrated, handheld thermometer. For the temperatures during cooling, three probes were affixed to in labeled cans such that their probe tips would be in the geometric center of the can, the latter filled with cheese and seamed. The data recorded is presented in Table 6.4.
6. The probes remained in the cans in the storeroom for at least 10 h before being removed and the data downloaded. The ambient temperature in the rooms was recorded.

Characterization of the thermal process

The thermal process applied to the production of pasteurized processed cheddar cheese at DIJL, while relying on the use of a high-temperature, short-time (HTST) cooking vessel characteristic of flash pasteurization systems, was not a typical flash pasteurization, HTST process. The product was rapidly heated to a temperature of 90°C in the cooker, but derived most of its lethality from the subsequent stages in which the high-residual temperature of the product delivered significantly greater lethality (or, more correctly, pasteurizing value or P-value) rather than during cooking (Table 6.4). This is because while most traditional process cheese pasteurization use processes, such as 85°C for 15 min, followed by rapid cooling and filling, delivering a P-value of approximately 340 pasteurization units (PU)/min, the Tastee cheese process did not involve rapid cooling. Consequently, the accumulated P-value during filling and, more importantly, cooling at >9100 PU far exceeded the total P-value of 3400–5100 PU delivered by a traditional process. The process is therefore best characterized as a *hot fill and hold process* as against the traditional *flash pasteurization* or *HTST process*.

In assessing this process, the *P*-value, expressed in PU is used instead of the F-value (the lethality) for pasteurization processes. The *P*-value is defined as the accumulated lethality (of a process) when $T_x = 60°C$ (140°F) and $z = 10C°$ (18F°), as indicated by the Institute for Thermal Processing Specialists (IFTPS, 2016), where T_x is the processing temperature and z is the z-value, which reflects the thermal resistance of microorganisms to different lethal temperatures.[h] Therefore, 1 PU is equivalent to 1 min at 60°C. The objective of pasteurization is to reduce the population of the target organism by 6 log, that is, by 10^{-6}. Industry has agreed by consensus a target figure of a minimum of 10 PU for *C. botulinum* Type E (Offley, 2016), but this figure is often doubled for an additional margin of safety, and so a *P*-value of 20 PU has traditionally been taken as being more than sufficient to assure safety of pasteurized products (Gordon, 2003). The FDA seeks to assure this in 21 CFR 133 by mandating time/ temperature combinations for milk-based products of 145°F for 30 min or 161°F for 15 s (Food and Drug Administration, 2012b).

Summary of thermal process considerations

The thermal process applied to the production of Tastee canned pasteurized processed cheese was characterized as a *hot fill and hold* process for a low-acid canned cheese product. It was sufficient to destroy all vegetative microorganisms of public health and economic significance of concern for pasteurized processed cheese products, such as these.

Formulation safety of Tastee canned processed cheese

In the manufacture of processed cheese and cheese spreads in the United States, controlling *C. botulinum* has been primarily the result of manipulating the pH, moisture (available water/a_w), and emulsifying salts [a combination of sodium chloride (NaCl) and phosphates] to deliver an environment hostile to its growth. Collectively, these ingredients and characteristics present sufficient hurdles for the organism so as to create an environment that is inhibitory for *C. botulinum* (Tanaka et al., 1979). In a further study of a variety of combinations of these factors using response surface modeling, Tanaka et al. (1986) were able to elaborate a set of principles for ensuring the safety of processed cheese from the growth and production of toxin in the product by *C. botulinum*. These principles, based on Tanaka et al. (1986)'s quadratic model depicted in Fig. 6.10, have become known as the *Tanaka Principles*, and are the bedrock of the processed cheese industry in guaranteeing food safety.

Tanaka et al. (1979) had previously shown that in the pasteurized process cheese spread packed in glass jars, the use of sodium phosphate as an emulsifier in combination with a moisture content of 54% or less, resulted in a substantial margin of safety against the growth of *C. botulinum*. This was despite having found

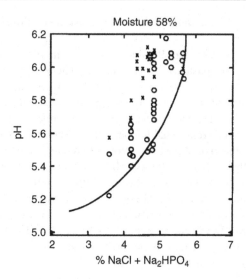

Figure 6.10 Response surface plot of pH against moisture and emulsifying salts content of pasteurized process cheese.
Source: Adapted from Tanaka, N., Traisman, E., Plantinga, P., Finn, L., Flom, W., Meske, L. and Guggisberg, J., 1986. Evaluation of factors involved in antibotulinal properties of pasteurized process cheese spreads. J. Food Prot. 49 (7), 526–531.

pockets of botulinum toxin production earlier in media with a high-protein concentration and a pH of less than 4.6, but which had localized areas characterized by precipitated protein and a high pH, in excess of 4.6 (Tanaka, 1982). Nevertheless, processed cheese product, with pH < 4.6, low moisture (<54%), and using a combination of sodium chloride and disodium phosphate, both of which inhibited botulinal toxin production with similar effectiveness, had a substantial margin of safety built in, as was expected (Tanaka et al., 1986). The greater inhibitory effect of sodium polyphosphate as against sodium citrate was confirmed by Ter Steeg et al. (1995), who also reported that while NaCl, lactic acid, and emulsifying salts aided stability, fat content was not a factor. This was confirmed by Ter Steeg and Cuppers (1995), who also used two other quadratic response models to confirm the Tanaka Principles on the effect of composition on safety of pasteurized process cheese. Glass et al. (1998) evaluated the impact of composition on other pathogens and found that while *S. aureus* survived it was unable to produce toxin, while numbers of *Salmonella*, *L. monocytogenes*, and *E. coli* O157:H7 declined over a 96-h storage period. Gordon et al. (2006, 2011)[i] also found similar survival patterns on multivac and pasteurized processed cheese. Consequently, products made with the ingredients indicated and which delivered the characteristics outlined in Table 6.5 would align well with the Tanaka model for "formula-safe" pasteurized processed cheese or cheese spread. This was the case for the product DIJL was currently manufacturing, which was therefore found to be fully compliant with the requirements of the model.

Table 6.5 **Comparison of the characteristics of Tastee cheese against the Tanaka requirements**

Parameters	Tanaka requirements	DIJL traditional product data
pH	5.8–6.2	5.6
Disodium phosphate	2.30–3.08%	2.1%
Sodium chloride	1.4%	1.7–1.9%
Total salts	3.75–4.20%	3.8–4.0%
Moisture	52–57%	44–47%
Nisin	—	0.05%
Sorbate	—	—

Product formulation: ensuring safety

The formulation for the production of Tastee Cheese was done with ingredients carefully selected to enhance the characteristics of the product and included cheddar cheese (>70%), water, disodium and trisodium phosphate, butter, citric acid, sodium chloride, annatto extract (an all-natural product), and nisin. These, collectively, along with the handling and production protocols implemented ensured compliance of the pasteurized processed cheddar cheese being made by DIJL with the requirements of the FDA. Specifically, to assure the safety of the canned pasteurized process cheese made by DIJL, it was necessary to ensure that DIJL was implementing handling and thermal processing requirements based a combination of factors that had to meet safe delivery. These combined safety "hurdles" allowed for the safe production of processed cheese without "sterilization." TSL also undertook a detailed assessment of DIJL's product. The assessment included that of a_w, for which it was shown that, although a contributor to safety, it was not sufficient by itself to assure food safety and was, therefore, not a good predictor of safety. In summary, TSL characterized DIJL's production of processed cheddar cheese in terms of its:

- moisture (including available moisture (water activity – a_w))
- pH, and
- total salts, such as:
  - NaCl and
  - phosphate-based emulsifiers.

This, combined with a very effective thermal process, which was capable of delivering a commercial sterile product, ensured that DIJL's Tastee canned pasteurized processed cheese was consistently safe and compliant with all regulatory requirements in all markets.

FRI's assessment of the data on the thermal process and product safety

Having fully understood the FDA requirements to get the import alert that had been imposed on the firm's product lifted, TSL then worked closely with the FRI to facilitate completion of the assessment of the canned processed cheese. FRI undertook detailed analyses of the information requested, as well as its own data gathered from assessment of the samples of the product sent to it at its request. The specific objective of the assessment, as indicated by FRI, was as follows: "To evaluate the safety of *Tastee Pasteurized Process Cheese Spread* produced by Dairy Industries (Jamaica) Limited, with particular attention given to the safety of the formulation with regards to controlling the growth and toxin production by *Clostridium botulinum*" (Glass, 2003). In doing the assessment, FRI requested and were supplied with the data and information outlined here:

1. The formulation of the product.
2. The list and details of analytical procedures used by DIJL, including:
   a. moisture,
   b. pH,
   c. NaCl, and
   d. water activity (a_w).
3. Analytical value for the product made over a 1-month period.
4. Details on the thermal process used.
5. Information on the nisin used as a bactericidal/antibotulinal agent in the product.

For all of these, FRI was provided with detailed information to allow it to undertake a thorough assessment of the product and process. The information provided to it is summarized here.

Formulation

The product formulation included cheddar cheese (>70%), water, disodium and trisodium phosphate, butter, citric acid, sodium chloride, annatto extract, and nisin. No other ingredients were added.

Analytical methods

All of the methods being used at the time were (and remain) Association of Official Analytical Chemists (AOAC) methods or those developed and validated by Fonterra Research Institute, the technical research center of one of DIJL's parent companies.

Analytical values

The details of mean analytical values as submitted to FRI are presented in Tables 6.2–6.5. Of importance in this regard were the a_w values, which were done by NP Analytical Laboratories for submission to FRI and showed that the results that were in the range of 0.947–0.955. Also provided for review by FRI were data on codes of product whose manufacture covered a period from recent production (July 2003) to expiration (>2 years postproduction), the analyses of which were done by Danisco International, the supplier of nisin to DIJL. Moistures ranged from 45.6% to 49.5% and the pH ranged 5.44–5.72.

FRI's summary of its assessment of the data

In assessing the data on moisture and pH for these products (including the expired product), Glass (2003) indicated that "…. Products that approach the upper limit for both moisture and pH have little or no margin of safety." In assessing the other data, Glass (2003) summarized the following information.

Analytical values

"DIJL provided a summary of analytical values for 19 batches of product produced on 27 March 2003 and 24 April 2003. Ranges for analyses included 41.2-42.4% moisture, pH 5.45-5.52, 1.84-1.92 NaCl, and 31.2-32.2% fat."

Thermal processing

"… The average cook temperature reaches 90°C for 0.16 minutes, with additional hold at >70°C for ≥6 minutes. The thermal processing described exceeds the minimum required by 21 CFR 133 to pasteurize process cheese and related products. Fill temperature at 70°C and maintaining the product at > 62°C for >16 minutes is sufficient to pasteurize the inside container surface. In addition, …. cooling to <62°C may take > 3 hours or longer."

Nisin

"Report by ….[j] Danisco International suggested that the addition of 11ppm nisin to the product will prevent botulinal growth and toxin production. Somers and Taylor[k] reported that 12.5 ppm nisin was required to be effective against *Clostridium botulinum* in process cheese spreads with 52.5% moisture, 3.9% totals salts, pH 5.9 and <25% fat. Recent research at the Food Research Institute indicates that the presence of fat reduces the antibotulinal activity of nisin in media." "The research by Somers and Taylor was completed in cheese spreads with approximately 20%. It is unclear the degree to which the high fat content of the Tastee cheese spread (26-32%) might reduce the efficacy of nisin in this product as compared to cheese spreads with 20% fat.". Consequently, as regards the use of nisin in the product, the conclusion was "Although it is unclear whether the high fat levels of the Tastee Cheese Spread have a significant adverse effect on the antibotulinal properties of nisin, continue to add 0.04% Nisaplin (10 ppm active nisin) unless additional formulation review and modifications are made."

All of these analyses and data, along with other factors, were considered by FRI and an opinion on the safety of the pasteurized canned processed cheddar cheese being exported to the United States from DIJL was prepared and delivered to the FDA and the firm (Glass, 2003). The recommendations are addressed further.

Resolution of the problem and regaining market access

The provision of the information on DIJL's product to the FDA/FRI and FRI's subsequent assessment of it allowed the FDA to make a definitive assessment on the compliance of the product. It indicated that as long as DIJL met the food safety requirements outlined by FRI in its report and followed the recommendations made (Glass, 2003), the product would be taken off the import alert and allowed reentry into the United States. FRI's recommendations follow.

FRI's recommendations

The FRI report (Glass, 2003) summarized its findings and recommendations as follows:

1) "Cheese spread is produced and packaged under Good Manufacturing Practices. It is recommended that the plant implement a well-designed HACCP program, or similar food safety system, for the production of this product."
2) "As part of safety system, records must document adequate thermal processing during production and hot-fill of product into cans to eliminate sensitive vegetative microorganisms. Current procedures used by DIJL exceed the requirements of 21 CFR Part 133.".... "Product temperatures within the can exceed 62°C for several hours, ensuring that vegetative microorganisms are inactivated on the interior can surface."
3) "Each batch of products must be analyzed using acceptable sampling procedures and standard methodologies for moisture, pH and NaCl to ensure compliance with production parameters discussed below."
4) "Records must document the addition of necessary amounts of sodium phosphate emulsifiers to provide a minimum level of total salts (defined as NaCl plus added anhydrous sodium phosphate emulsifiers) and the addition of 0.04% Nisaplin (equivalent to 10.0 ppm active nisin)."
5) Production levels must be maintained as below to provide an appropriate margin on safety.
   a) Target 44-47% moisture, with no release of product to US if moisture exceeds 47.5%
   b) Target pH 5.4-5.6, with no release of product to US if equilibrated pH exceeds 5.70
   c) Target 3.8% or greater total salts (defined as analyzed % NaCl plus % added disodium and trisodium phosphate), with no release of product to the US if analyzed NaCl is <1.67% or if calculated total salts is <3.75%

With regards to the last and, perhaps most critical of the recommendations, Glass (2003) advised that "should a batch of product fall outside of the specifications for moisture, pH or salts, product should be held until reviewed by process authority. Multiple samples should be evaluated for water activity. Product with $a_w > 0.96$ (at 25°C) should not be released for sale unless written variance based on pH and salts is given by process authority."

Implementation of the recommendations from FRI and regaining of the market access

Prior to the assessment of the production of DIJL's canned pasteurized processed cheddar cheese, the activities that were being undertaken as part of monitoring and process control were:

• measuring ingredients,
• monitoring cooking temperature,
• monitoring moisture,
• monitoring pH,

- monitoring fat, and
- monitoring salt (NaCl).

As an outcome of the report, several critical activities that were central to the FDA's required controls and regulations were identified and became the focus of the process. These were:

- monitoring of initial temperature,
- monitoring of fill temperature (FT),
- continuous monitoring of temperatures (<15 min),
- monitoring and recording of pH,
- checking and recording of can seams (>3 times daily),
- monitoring NaCl and total salts (disodium and trisodium phosphate and sodium chloride),
- monitoring and control of moisture levels, and
- monitoring residual nisin levels.

In addition to these, DIJL was required to ensure that its facility and products were in full compliance with the FDA requirements, including

1. Registering the factory with FDA under both the Bioterrorism Act and mandatory food facility registration.
2. Filing the process for Tastee's canned pasteurized processed cheddar cheese with FDA's LACF Division.
3. Agreeing to implement a compliant HACCP program.

DIJL implemented all of the recommendations of the FRI report, in addition to maintaining its own stringent controls that were in place previously. The FDA verified that all was in place and, in December 2003, lifted the import alert on *Tastee Pasteurized Processed Cheddar Cheese Spread*, allowing it to recommence exports to the US market.

Conclusions

Traditional products, such as the canned pasteurized processed cheese product discussed in this case study can face market access interruptions if their production is not backed by appropriate science. These products may rely on a variety of nontraditional "hurdles" or combined methods, as in this case, based on concepts, such as the Tanaka Principles to ensure their safety. In cases, such as these the approaches taken and the information presented can be useful in leading toward a solution to any food safety challenges that may arise for dairy-based and other similar products.

Case study 2: attaining compliance of canned pasteurized processed cheese with Canadian requirements

Background

DIJL had been exporting the same Tastee canned pasteurized processed cheddar cheese to Canada until the mid-2000s. At that time, the Canadian authorities advised that the product, was noncompliant with Canadian statutes and could no longer be imported into Canada. The noncompliance was due to the use of nisin in the product, an additive that, while allowed in the United States, was prohibited in dairy products in Canada. DIJL sought to address the problem by substituting potassium sorbate for nisin in its Tastee canned processed cheese spread. This created another challenge, as when asked to provide evidence that the processed cheese product remained safe and compliant with the original science that facilitated its entry into the United States, the firm was unable to do so, neither having done the studies, nor able to find research specifically supporting the efficacy of the change that was made. DIJL contracted TSL to undertake the technical work necessary to facilitate the export of canned processed cheese made using sorbate as an antimicrobial agent to the Canadian market. This required, among other things, a validation study to show that the use of potassium sorbate in the product resulted in a product that not only complied with Canadian requirements but also met DIJL's marketing objectives. This case study documents the approach taken, providing DIJL with the supporting information and documentation it required to replace the vacuum packaged cheese in the Canadian market with its canned pasteurized processed cheese, which is more favored by its market.

Approach taken

The approach taken involved a comprehensive assessment of the regulations govern-ing the product, a review of the Tanaka Principles for dairy products, including updates on recent work, a detailed thermal process assessment to compare the suitability of the process against that being used for the traditional product, and an assessment of the prac-tices used at the facility for canned cheese production.[1] The formulation was reviewed to establish conformance to the Tanaka Principle requirements. A review of the use of potassium and other sorbates in dairy products and their antimicrobial efficacy, as well as an assessment of the robustness of the product when challenged by major pathogens, and determination of the shelf life of the product were also done to validate the use of sorbate to replace nisin in the processed cheese. The challenge study involved the use of *Clostridium* sp., *L. monocytogenes*, and *Geobacillus stearothermophilus*, among other organisms. Data on Tastee cheese made with sorbate over the past 2–3 years was also reviewed and analyzed. A detailed review of thermal process/can seam/cooling data was undertaken to verify the efficacy of the process being applied in delivering a safe product.

Specific aspects of the Canadian regulations affecting canned Tastee processed cheese (Canadian formulation)

A detailed review of the pertinent Canadian laws and regulations that could impact the formulation, processing, production, export, and sale of the product [canned Tastee processed cheese (Canadian formulation)] being contemplated for export to Canada was done. This covered the use of sorbates in dairy products, specifically processed cheese, as well as the thermal process requirements, permitted additives, terms of trade, and import and compositional aspects of the product, which are controlled under Canadian statute. From this exercise, the relevant sections of the laws and regulations were evaluated and this evaluation documented. The information gathered, as well as implications of these and recommendations are summarized in the information presented in the further sections.

Tariff rate quotas

First come first served (FCFS) tariff rate quotas (TRQs) are administered by the Department of Foreign Affairs and Canadian Border Services Agency, and are administered for margarine, wheat, wheat products, barley, barley products, cut roses from Israel, frozen pork from the European Union, etc. There are also TRQs that are subject to allocations and these are administered by the Department of Foreign Affairs. This was the category of TRQs that was applicable to cheese, among other products, and hence, this was where the import status of the product and the quantities allowed for import would fall, all other regulatory criteria being met.

Permitted preservatives

The list of food additives, including preservatives that are permitted for use for foods to be consumed in Canada, was available and provided by the Canadian Government on the Health Canada website at http://www.hc-sc.gc.ca/fn-an/securit/addit/list/11-preserv-conserv-eng.php. This information and other sources were reviewed and it was concluded that potassium sorbate and sorbic acid are allowed as permitted additives to foods consumed in Canada at their stipulated limits in particular foods.

In the Canadian regulations, sorbates are approved for use in dried meat, baked goods, jellies, jams, and syrups, among other products and in cheese, including processed cheese. A variety of sorbates are approved for use in Canada, including calcium sorbate (203),[m] potassium sorbate (202), sodium sorbate (202), and sorbic acid (200). The maximum level for use of sorbic acid/sorbate indicated for processed cheese is listed as 3000 (ppm).[n] For unstandardized processed cheese products, a maximum level of 3000 ppm is also applicable. This is the category into which Tastee canned processed cheese fell. If calcium sorbate or potassium sorbate is used, the total concentration must not exceed 3000 ppm, calculated as sorbic acid. This is in line with the Codex Alimentarius, which also listed 3000-mg/kg (ppm) sorbates as the maximum level for sorbates as a food additive in processed cheese.[o]

Composition of processed cheese

The minimum milk fat allowed in processed cheddar cheese by the Canadian standards is no more than 3% below that present in natural cheddar cheese. This means that at 34%, DIJL)'s Canadian cheese should have no less than 31% milk fat. Batches over the last 3 years have varied between 30% and 32%. For water content (moisture) a maximum of 44% is allowed. DIJL's product met these requirements.

Summary of Canadian regulatory compliance and recommendations

In summary, review of the Canadian regulations advised the following:

- DIJL could export canned cheese to Canada under the FCFS TRQ system, as is being done for the vacuum packed product.

- *Potassium sorbate*, the replacement being used for *nisin* in the processed cheese, can be used at a limit of 3000 ppm (or 0.3%), maximum. DIJL currently uses 0.13%, which is well within the limit.
- Processed cheese fat level for DIJL's product could not fall below the minimum of 31% allowed, based on its existing formulation. Fat levels would, therefore, need to be increased to assure compliance.
- For moisture, the maximum water content allowed in Canada for this product is 44%, which comfortably met DIJL's product, requiring no further modification.

The use of sorbates and sorbic acid

Sorbic acid and its salts are used as a preservative against fungi, yeast, and bacteria. They are polyunsaturated fatty acids that do not affect the taste or the color of the food product. Sorbates are most effective at pH 5.5–6.0. Sorbic acid is slightly soluble in water and completely soluble in alcohol. Potassium sorbate on the other hand, is more soluble in water and slightly soluble in alcohol. Typical level in products, such as processed cheese, is 0.2–0.3%. Sorbates work through the transmission of some of the undissociated acids in the external (food) medium into the cytoplasm of the cell of the targeted microorganism. As the cytoplasmic pH is close to neutral, weak acids, such as the sorbates, dissociate and decrease the pH of the cytoplasm, resulting in bacteriostatic/bactericidal effects.

Compliance of the traditional Tastee cheese formulation with the Tanaka Principles

The traditional canned Tastee processed cheddar cheese had been formulated and produced since 2003 to fully comply with the Tanaka Principles as outlined previously and as advised by FRI (Glass, 2003). This, as was discussed in the previous case study, included:

- A target moisture of 44–47% with a maximum moisture of 47.5%.
- A target pH of 5.4–5.6, with a maximum equilibrated pH of 5.70.
- A target total salt content (defined as analyzed % NaCl plus % added anhydrous disodium and trisodium phosphates) of 3.8% or greater.

As long as these specific requirements were met and the thermal process prescribed by DIJL's Process Authority followed, the product (which was formulated with nisin to provide a further barrier against the outgrowth of *C. botulinum*), would remain commercially sterile for the duration of its commercial journey. It would also not present a risk of causing foodborne illness to consumers and deliver the full shelf life expected (18 months to 2 years, minimum).

Typical data on canned processed cheese product (Canadian formulation)

The typical composition of the modified blend (formulation) of Tastee cheese made for the Canadian market is shown in Table 6.6. This compared favorably with the requirements for processed cheese to be shelf stable and safe based on the Tanaka

Table 6.6 **Typical chemical composition of Tastee processed cheddar cheese (Canadian blend)**

Parameters		Range	Laboratory analysis
Precook	Moisture	36–39%	*37*
	pH	5.45–5.54	*5.46*
	Salt	≥1.8%	*2.08*
	Fat	31–32%	*34*
Postcook	Moisture	41–44%	*42.3*
	pH	5.4–5.7	*5.5*
	Salt	1.67–2.00%	*1.84*
	Fat	31–33%	*31.0*
	Total Salt	≥3.75%	*3.77*

requirements (Table 6.7). The analytical data for product made with this formulation over a period of 3 years for the exiting vacuum packed variant of the product also showed consistent analytical results. The product consistently had a moisture of 44% (maximum), pH of 5.6 (maximum), and total salts of 3.8% (minimum), which was in full compliance with the requirements; therefore, allowing a level of assurance of quality, safety, and compliance for the reformulated product. The data showed, therefore, that the Tastee processed cheddar cheese produced for Canada was safe for consumption, all critical factors having been met. As such, the product with the added sorbate should be able to deliver a similar level of shelf stability and microbial inhibition based strictly on the Tanaka Principles, without the need to involve the bacteriostatic effect of sorbate.

Production process and thermal process adequacy

As a part of the assessment of the conformance of the manufacture of this product to the requirements, an assessment was done of the overall process of making canned Tastee processed cheddar cheese for export to Canada. This involved a review of the

Table 6.7 **Comparison of Tastee cheese varieties against the Tanaka requirements**

Parameters	Tanaka requirements	DIJL traditional product data (regular Tastee cheese)	DIJL product data (Canadian Tastee cheese)
pH	5.8–6.2	5.6	5.5
Disodium phosphate	2.30–3.08%	2.1%	2.16%
Sodium chloride	1.4%	1.7–1.9%	1.6–1.9%
Total salts	3.75–4.20%	3.8–4.0%	3.67–4.16%
Moisture	52–57%	44–47%	41–44%
Nisin	—	0.05%	—
Sorbate	—	—	0.13%

handling of the raw materials process flow and HACCP plan, as well as the direct observation of the process. Having completed this assessment, a detailed assessment was then done of the efficacy and adequacy of the thermal process applied, including the P-value attributable to each stage of the process and the overall impact of the process on any vegetative cells of pathogens that may be present in the product. This information, as well as the accompanying data that was reviewed is presented here.

Thermal process adequacy assessment

Methodology

A detailed review of thermal process/can seam/cooling data to show comparative safety of the product was undertaken. The objective of this assessment was to validate the effectiveness of the process being used to manufacture the product and demonstrate that it delivered a safe product. If this proved not to be so, the data gathered was expected to facilitate the development of an appropriate process that would achieve the objective of delivering a safe, high-quality product. Two 2000-kg batches of Canadian processed cheddar cheese, canned in 2.2-kg cans were prepared particularly for the assessment of the adequacy of the thermal process. The temperature throughout the process at was checked at five points in the process: the auger cart, divert valve, exit from flash (deaeration) vessel, filling, and during cooling (see process described in case study 1). In addition, where relevant (e.g., as in the warehouse during cooling), the ambient temperature was also monitored and recorded. The protocol used for the thermal process study was as described in case study 1, with DataTrace MPIII probes being used instead of MPII probes (Fig. 6.11).

Findings and implications

The data captured by DIJL's internal recording systems including process (cooking) temperature are presented in the Fig. 6.8, and the placement of cases containing the DataTrace MPIII probes in the warehouse and their locations on pallets during the cooling process are presented in Fig. 6.12. The test pallet was located as shown (Fig. 6.9), with no pallets above it. The data for temperatures captured throughout the process are presented in Tables 6.8–6.10. The outcome of the study of the current process being used shows that the product moves from a temperature of about 18.2°C at the augur cart, through cooking at a mean temperature of 93.3°C, to filling at a minimum of 72.3°C (Tables 6.8–6.10; Fig. 6.8). Both the probes and the calibrated thermometer show FTs in the range of 73°C on average, with the former recording temperatures as low as 72.3°C (Table 6.9). The FT is critical because it directly determines the lethality that will be accumulated during cooling (Gordon, 2003), and the data show that cooling begins the moment cooking is completed and by the time the product exits the flash vessel, before filling, it is already cooled by 11.1°C (Table 6.10). The effectiveness of the thermal process applied to this product, as already mentioned, was previously determined to be largely due to the accumulated lethality as a result of the cooling step (Gordon, 2003), this being derived mainly from the FT.

  The various pasteurizing values (PVs) associated with the processing steps for traditional Tastee processed cheddar cheese, taken from Gordon (2003) are presented

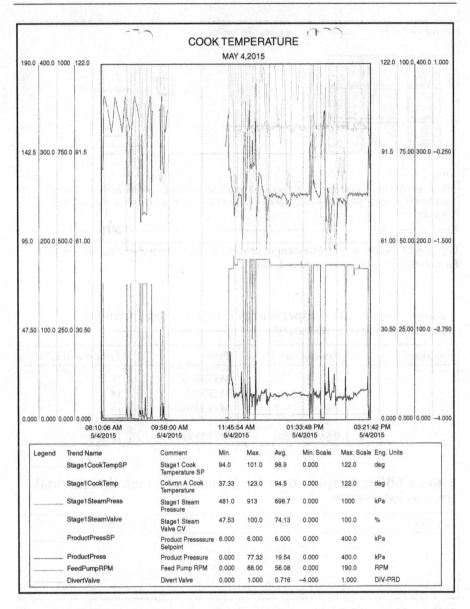

Figure 6.11 Cook temperature for Canadian formulation processed cheese on May 4, 2015.
Source: André Gordon, 2003.

in Table 6.11. The outcome of this study, although for a different formulation, shows congruence between the two, with PVs being derived in a similar manner and quanta (Table 6.11 vs. Table 6.12). What is noteworthy, however, is that despite higher cook, post flash vessel, and minimum FTs, the accumulated PV during cooling is less than in the previous study. This may be influenced by the differences in formulation, which

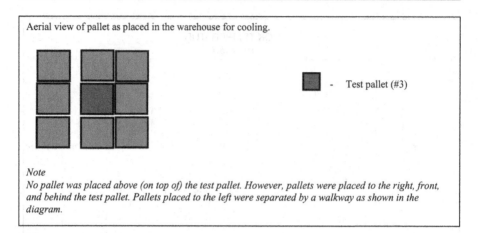

Figure 6.12 Location and placement of probes on pallets during cooling for Canadian formulation processed cheese May 4, 2015.

Table 6.8 **Auger cart temperatures for Canadian formulation thermal process assessment**

Samples	Time of day	Probe	Temperature (°C)
1	9.23	M3T39510	18.2
2	9.48	M3T39504	18.3
3	9.49	M3T39508	18.3
4	9.49	M3T39501	18.4
5	9.56	M3T39506	18.5

Table 6.9 **Filling temperatures for Canadian formulation thermal process assessment**

Samples	Time of day	Temperature (°C)
1	9:26	72.7
2	9:50	73.3
3	9:51	76.0
4	9:52	72.3

caused the Canadian formulation to be less efficient at holding heat or, alternatively (and more likely), it could be influenced by the longer times taken to fill the product, lower FT's (on average) than the previous study, and hence, a lower accumulated PV. In both cases, *the accumulated PV is far in excess of the 20 PU required to achieve pasteurization of processed cheese* (Gordon, 2003). *Consequently, the thermal process used for this product should be sufficient to deliver a commercially sterile product with no viable vegetative cells (live organisms) that gives a shelf life of at least 2 years.*

Table 6.10 Cooker and cooling temperatures for Canadian formulation thermal process assessment

Samples	Heating process	PU	Time of day	Probe	Temperature (°C)
1	Cooking	—	9.27	M3T39510	93.33
2	Flash vessel	—	9.46	M3T39508	82.22
3	Cooling	7724.00	9.48	M3T39500	73.5 (Fill)
4	Cooling	9620.00	9.52	M3T39501	76.7 (Fill)
5	Cooling	7866.50	9.58	M3T39506	73.9 (Fill)

Table 6.11 Original process and thermal parameters derived for canned processed Tastee cheese (traditional formulation)

Stage of process	Average temperature (°C)	Average time (min)	Pasteurizing value (PU)
Blending	19.5	—	—
Cooking	90	0.16	152.1
Holding	72	10	142.3
Filling	70	16	227.7
Cooling*	<62	>240	9100

*Maximum lethality achieved by the process.

Table 6.12 Thermal processing parameters for canned processed Tastee cheese (Canadian formulation)

Stage of process	Average temperature (°C)	Average time (min)	Pasteurizing value (PU)
Blending	18.2	—	—
Cooking	93.2	0.16	175.8
Holding	82.0	15	162.5
Filling	72.3	20	202.3
Cooling*	<62	>240	8400

*Maximum lethality achieved by the process.

It should be noted, however, that during the trials and subsequent simulations (see section "Challenge Study: Survival of Selected Pathogens in Tastee Canadian Formulation Processed Cheese"), it was noted that relatively low FTs could result in an understerilization of cans and lids and insufficient destruction of the vegetative cells of selected pathogens, if present at high levels. These findings indicate that FTs are even more critical for this formulation with a relatively low sorbate content of 0.13% than for the traditional formulation made with nisin. This, therefore, suggest that greater caution needed to be taken with this product and hence, a FT below 76°C was not recommended as this may predispose the product to spoilage/reduced safety. Both sulfite-reducing clostridia (SRC) (including *C. perfringens*, and potentially also

C. botulinum) and *Listeria* may survive in product filled at lower temperatures and put the product and consumers at risk, depending on their initial loads.

Shelf life

An accelerated shelf life study was done on Tastee processed cheese formulated for the Canadian market (i.e., with sorbate). The product evaluated [Tastee Canadian formulation processed cheese (TCFPC)] was manufactured by DIJL over time, including product made under TSL's guidance in April 2015. The samples were placed in storage at different temperatures and were assessed organoleptically, chemically, and microbiologically. The approach used and the resulting data from the study are presented further. These were used to determine the acceptable shelf life for the product to guide not only the declaration of "best before" on the cans but also practice in the commercial distribution operations in Canada.

Approach and methodology

The approach used in this study was the one used for Tastee processed cheese spread (Gordon et al., 2011), which was previous developed by Gordon et al. (2006) using the approach recommended by Vanderzant and Splittstoesser (2001) and Roberts et al. (1986). The only modification was that in this study, the products were incubated in the cans to simulate conditions of the Canadian marketplace where they would be sold. All products were manufactured at DIJL using the specific formulation for Tastee Canadian processed cheese (TCPC) as would be the normal commercial practice and packaging, prior to being selected for use in the study. Product evaluated represented four production dates December 2013, February and December 2014, and May 2015, the latter being made under TSL's guidance.

A total of 10 units for each date code were used for the assessment, with product being taken for sampling over the period of the trial, which lasted for 56 days. Units were incubated at 4, 25, and 35°C, representing refrigerated (and therefore stable conditions), ambient (room temperature in Canada), and accelerated deterioration conditions, respectively. Samples were taken from storage at the relevant temperatures for sampling, as required throughout the duration of the study. The products were evaluated for their sensory (organoleptic and olfactory) characteristics, pH, and the presence of selected indicator and pathogenic organisms over the duration of the period of incubation. The organism assayed for were total aerobic plate count (TAPC), total anaerobic spore count (TASC), and SRC.

Findings and implications

The shelf life (θ) of the TCFPC was determined using three storage conditions for each the set of samples, as indicated previously. These represented stable conditions under which the product was not expected to change dramatically (4°C), conditions typical of the expected temperatures under normal distribution conditions in Canada (25°C), and an abuse temperature that was also used to accelerate any potential deterioration of the product. This involved storage at an elevated warehouse temperature (35°C) for the duration of the study, and facilitated a prediction of the shelf life of the product over an extended period of storage Fig. 6.13.

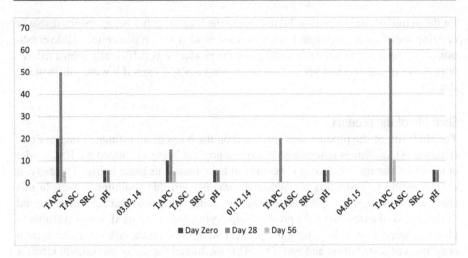

Figure 6.13 Selected shelf life parameters for Tastee Canadian formulation processed cheese (TCFPC) stored at 25°C for 56 days. *TAPC*, Total aerobic plate count; *TASC*, total anaerobic spore count, *SRC*, sulfite-reducing clostridia.

The findings

The results of holding at the normal distribution temperatures in Canada of 25°C (i.e., the temperature the product is likely to be held at during the spring, summer, and early autumn in this temperate country) on selected microbiological and chemical characteristics of the product are shown in Fig. 6.13. The data show that there was little change in the key characteristic of pH (and conversely, the acidity) over the period of storage, indicating an acceptable level of stability of the chemical characteristics of the product. Data also indicated that all of the different batches of cheese (02.12.13, 03.02.14, 01.12.14, and 04.05.15) evaluated at all temperatures showed this same stability and also showed very little variation in their pHs, indicating the stability not only of the formulation, but also of the production process for TCPC. The pH of all products ranged from 5.47 to 5.59 over the period, with the batch produced in 2015 having the highest pH (5.56 on average). The data for the stability of the product stored at 4 and 35°C (not shown), follow a similar pattern to that presented in Fig. 6.10.

The initial microbiological analyses on all products confirmed the original findings at DIJL and found all to be microbiologically acceptable with low TAPC (<100 CFU/g), no viable anaerobic spores, and also no viable SRS (Fig. 6.10). The absence of sporeformers was consistent throughout the course of the shelf life study, indicating that the formulation for the TCPC is robust enough to prevent the outgrowth of any spores that may be present, particularly *C. perfringens* or *C. botulinum* spores, should they be present. This also confirmed the findings and recommendations of the thermal process assessment work done in a companion study (Gordon et al., 2015b), which also showed that the product was safe and stable for distribution and sale.

The sensory assessment of the product indicated that while there was a slight increase in the sharpness of the product held at the elevated (abuse) temperature of 35°C for the period of the study, this did not rise to the level of either being objectionable or of being significantly different from the cheese held at 4°C, representing stable conditions. This further validated the finding that the product was sufficiently robust to have only minor organoleptic deterioration in the marketplace, even if it were substantially abused.

Shelf life of the product

The shelf life of the product was predicted on the basis of an estimated rate of deterioration (Q_{10}). This is determinable when a product being evaluated by 10°C under at least two storage conditions is deemed to have passed its useable life. Typically, it is challenging to make this prediction accurately until this condition is met. In this study, this condition was not met because the product remained viable up until the end of the period of assessment. As products of varying ages were used, it was possible to give a projection as to how long the product would last on the market under normal conditions of distribution and handling. This facilitated the use of the growth kinetics of the organisms selected, as well as the rate of deterioration of other characteristics, and an assessment of a matrix of possibilities for projecting a minimum shelf life for the product. The predicted shelf life is based on these considerations and the observations made on the TCPC, supported by calculations based on the projected rate of deterioration.

Based on these findings, the calculations, and the assumptions governing them, the product had an estimated shelf life of approximately *960 days* (*2 years, 7 month*) at 4°C (θ_4), with a θ_{35} of at least *275 days* at the elevated temperature (35°C). The shelf life at 25°C (θ_{25}) is projected to be *830 days* (*2 years, 7 months*). A conservative Q_{10} for DIJL's TCPC, based on changes in the pH and other chemical and organoleptic factors under the most abusive conditions assessed (35°C) was *3.02*.

It is therefore recommended that a *maximum* shelf life (θ) of *790 days* (*or approximately 2 years and 2 months*) at ambient distribution temperatures (25°C) be used. This should give a significant margin of safety for the product during distribution.

Summary

The results of the accelerated shelf life study showed that canned TCFPC was a stable product that could be safely be distributed in Canada under the normal conditions of trade. Under these conditions, a shelf life in excess of *2 years* was to be expected.

Challenge study: survival of selected pathogens in Tastee Canadian formulation processed cheese

Introduction

This aspect of the project sought to determine whether or not the formulation and process being applied were, collectively, sufficient to ensure that Tastee canned processed cheese (Canadian formulation) would be safe as produced. These studies also sought to determine the robustness of the formulation and the process applied in resisting the growth of selected pathogens, should they be present. It was undertaken with the commercially produced product that was then used for the trials, which was inoculated with selected pathogens. These samples were then held under varying conditions and assessed over time to determine whether the formulated product would support the growth of specific pathogens or provide an inhospitable environment for growth.

Approach and methodology

Commercially produced Tastee cheese (Canadian formulation) was made at DIJL, collected through the divert valve and transported under conditions that kept the product hot to an offsite laboratory.[p] The cheese was filled into sterile 250-g cans and inoculated with known concentrations of selected pathogens. Cans were seamed under nitrogen to create a much stronger anaerobic environment than would typically be the case in commercially produced Tastee cheese, which is filled at ambient temperature and sealed in the presence of air, creating a much less favorable environment for the growth of anaerobic or facultatively anaerobic pathogens than in this study.[q] Positive and negative controls were used. The pathogens used were *L. monocytogenes*, SRC (*C. perfringens*), and *S. aureus. Geobacillus stearothermophilus* was also used as an inoculant, as this organism is both thermophilic and fairly robust and so grows in conditions where *C. perfringens* may not, leading to spoilage of the product. Products were incubated at 4, 25, 35, and 55°C and then enumerated for the selected organisms according to standard protocols.[r] Day 0 specimens were also assessed for TAPC, total aerobic spore count, and TASC to establish a baseline and validity of the procedure. Environmental samples were also done to provide a control for random contamination.

Findings and implications

The results of the challenge study and survivability assessment for the Canadian formulation Tastee cheese are presented in this section. Environmental plates showed no presence of unusual potential contaminants and positive and negative controls were unremarkable. The results for controls evaluated after 5 days of incubation at room temperature are presented in Table 6.13. The results showed that neither *G. stearothermophilus* nor *S. aureus* were able to survive in the positive control samples, canned under nitrogen. They also did not survive in any of the other samples into which they were introduced (Tables 6.13 and 6.14). *Listeria* and SRC survived and thrived in all samples at all temperatures, except for 55°C (Table 6.14) where *Listeria* did not survive and SRC struggled.

The survival of *Listeria* in all other conditions is consistent with the findings of work done previously with processed cheese (Gordon et al., 2006, 2011). However, while both *Listeria* and *S. aureus* were found to survive well in vacuum-packaged pasteurized processed cheese in both studies, the latter did not do so in this study. SRC survived in this study (Tables 6.13 and 6.14), but it did not in vacuum-packaged pasteurized processed cheese (Gordon et al., 2006), which, though also an anaerobic environment, was a different formulation from what was used here.

Table 6.13 Challenges and survivability data for selected pathogens in Tastee canned processed cheese, Canadian formulation: controls

Organisms	Can #			
	B1	33	37	39
G. stearothermophilus	<5	<5	<5	<5
S. aureus	<10	<10	<10	<10
L. monocytogenes	+ve	+ve	+ve	+ve
SRC	16,000	58,000	57,800	84,000
TAPC	<5	<5	<5	<5
Total anaerobic plate count	20,000	150,000	180,000	120,000

Table 6.14 Challenges and survivability data for selected pathogens in Tastee canned processed cheese, Canadian formulation: Day 25

Organisms	Temperature of storage (°C)					
	4	4	25	25	55	55
G. stearothermophilus	<5	<5	<5	<5	<5	<5
S. aureus	<10	<10	<10	<10	<10	<10
L. monocytogenes	+ve	+ve	+ve	+ve	−ve	−ve
SRC	8,800	11,000	4,200	24,000	50	<10

An important feature of this study was the temperature at which the canned processed cheese was filled. For most of the cans, the temperature at filling ranged from 58 to 70°C, which was under the recommended FT for this product. Under these conditions, both *Listeria* and SRC survived. This means that these FTs, combined with the holding time and the formulation, were not enough to prevent the survival and outgrowth of these pathogens when inoculated at the very high titers used in the study. This suggests:

1. that lower FTs in the commercial process, should these organisms be present in large numbers or be able to survive cooking, may put the product at risk and
2. the formulation is not sufficiently robust to prevent the outgrowth of large numbers of these organisms, if present.

The inability of the heat supplied by lower FTs to generate enough lethality is an indication that great robustness was possibly needed with regard to the process for this product. The fact that protection by the Tanaka Principle–based formulation did not seem to be sufficiently effective under these conditions further supports this observation. It appeared the presence of sorbate at the relatively low levels used, while possibly effective against the other organisms, was not against *Listeria* and clostridia. Both SRC (including *C. perfringens*, and potentially also *C. botulinum*) and *Listeria* survived well in product filled at temperatures below 70°C and were not inhibited by the sorbate at level used.

Conclusions and recommendations

All of the data on this product, from these studies and previous work, indicated that the combination of the thermal process used, along with the formulation should be sufficient to deliver a commercially sterile product with no viable vegetative cells (live organisms) that gives a shelf life of at least 2 years. However, the findings of the challenge study indicated that relatively low FTs could result in an understerilization of the can and its contents and insufficient destruction of the vegetative cells of selected pathogens, if present at very high levels. Further, the relatively low level of sorbate used (0.13% sorbate) was not sufficiently effective against *Listeria* and clostridia if these organisms were to be present at very high levels in this product. These findings indicated that FTs below 78°C (76°C minimum) for the formulation existing at the time could predispose the product to spoilage/reduced safety.

Recommendations

The following recommendations were made:

- This particular product, as formulated, should not be filled below 77°C (i.e., 76°C would be the critical limit at this critical control point), with the operating control limit being 78°C. Controlling the cooking temperature (target 92°C and above) and minimizing the holding time prior to filling are important.
- The optimal conditions that comply with the requirements of the Tanaka Principles: moisture 44% (maximum), pH 5.6 (maximum), and total salts 3.8% (minimum) must be maintained for this product and, where possible, more robust conditions sought.
- The sorbate level should be increased to the maximum allowable under the regulations, which also do not affect the flavor profile of the product.
- DIJL needs to ensure that scrupulous care is taken to eliminate SRC, in particular, from the environment when this product is being made.
- Other recommendations made in other aspects of this work should be considered to enhance the safety of the product.

The recommended adjustments to the formulation and controls were made and the product as currently made has been consistently free from organisms of public health or economic significance, and has a shelf life of in excess of 24 months. DIJL therefore was able to produce a viable product to take advantage of an export opportunity, having implemented the recommended adjustments to the product and process critical control points. The product now meets all of the requirements to be imported into Canada and has sufficient safety barriers to be sold in the marketplace.

Endnotes

[a]See the website of retailers: http://www.canned-cheese.com/.

[b]This description arises from the US requirement for the name of the product to reflect their regulations for Standards of Identity encapsulated in their regulations (Food and Drug Administration, 2015e).

[c]It was neither unusual at that time, nor is it even today for many traditional products that have been made for years to not have been fully scientifically characterized, even when they are

being managed by best-in-class quality systems [Dairy Industries Jamaica Limited (DIJL) was ISO 9001 certified for many years at the time]. This scientific characterization and validation of process and other preventive controls is being mandated by the FSMA through the Preventive Controls Final Rule.

[d]Dr. André Gordon, the author and Technological Solutions Limited's (TSL) principal consultant.

[e]"Sterilized" in this context means processed at a temperature and for such a time as to destroy all viable vegetative cells and spores of microorganisms of public health significance.

[f]The discussions were facilitated by Dr. Dennis Dignan (now deceased) and Mr. Steve Spinak (FDA, retired).

[g]Dr. Glass is a food safety specialist, with an extensive body of work on food safety and is widely regarded as the leading expert on pasteurized processed cheese safety. She is a Distinguished Scientist and the Associate Director of the Food Research Institute (FRI) at the University of Wisconsin, Madison, Wisconsin, United States.

[h]The z-value is defined as the temperature change required to change the D value by a factor of 10. The D value is a measure of the heat resistance of a microorganism. It is the time in minutes at a given temperature required to destroy 1 log cycle (90%) of the target microorganism (Heyliger, 2012).

[i]These studies were done as a part of the validation of the safety of Tastee brand process cheese to allow for registration and thermal process filing with the FDA and export to Canada, respectively.

[j]Dr. Joss Delve-Broughton, Danisco's specialist on this matter at the time.

[k]Somers and Taylor (1987) did a study on the antibotulinal effectiveness of nisin in pasteurized process cheese spread.

[l]The work was led by Dr. André Gordon, TSL's Thermal Process Authority supported by TSL's other resources.

[m]203 is the number specifically assigned to calcium sorbate in the Canadian regulations (similar to E-numbers in the EU regulations).

[n]Sections B.08.033, B.08.034, B.08.035, B.08.037, B.08.038, B.08.039, B.08.040, B.08.041, B.08.041.1, B.08.041.2, B.08.041.3, B.08.041.4, B.08.041.5, B.08.041.6, B.08.041.7, and B.08.041.8) of the Canadian *Food and Drug Act and Regulations*.

[o]http://www.codexalimentarius.net/gsfaonline/foods/details.html;jsessionid=7946DE4E5C38 A77928A34D3B6817ED97?id=30

[p]TSL's laboratory.

[q]The objective was to create an environment that was very favorable to the survival and growth of pathogens to test the limits of the product as regarded by its ability to prevent their growth.

[r]This was done by the standard protocols used in TSL's ISO 17025 accredited laboratory.

Case study: application of appropriate technologies to improve the quality and safety of coconut water

7

A. Gordon*, J. Jackson**
*Technological Solutions Limited, Kingston, Jamaica; **Independent Consultant, Okemos, MI, United States

Chapter Outline

Food Safety and Quality Systems in Developing Countries. http://dx.doi.org/10.1016/B978-0-12-801226-0.00007-4

Introduction

The coconut

Coconut (*Cocos nucifera* L.) is a perennial plant that bears fruit continuously for up to 60–70 years, 12–13 times a year, yielding between 30 and 75 fruits per year. The fruit, in the form of a nut (Fig. 7.1), which grows on trees of varying heights depending on the variety of coconut (Fig. 7.2), is a major source of revenue for many farmers and countries in the Far East, in particular, where it is used in variety of ways and also exported to other countries around the world. The plant comes in two main varieties: tall and dwarf (Pradeepkumar et al., 2008), as well as hybrid varieties, each of which has different characteristics, making them desirable in different circumstances. The tall tree such as the Sri Lanka or West African tall can grow up to 25–30 m while the dwarf, such as the Malayan dwarf typically does not exceed 4–8 m in height. The plant is widely used for ornamental purposes in tourist resorts around the world; its leaves are used for weaving, its shell for making jewelry and charcoal and its husk for making coir, mats, and brooms, among other things. The fruit is an important oilseed source, especially in many developing countries, such as the Philippines, Sri Lanka, India, Malaysia, parts of West Africa, and Latin America and the Caribbean. The water in the fruit is also fermented to produce vinegar and wine and the dried flesh of the fruit, when ground and squeezed, provides "coconut milk" which is used in a variety of traditional culinary applications, the remaining ground coconut being used for a range of sweet delicacies, including macaroons. Desiccated coconut is in high demand the world over, with more than US $800 million in global exports led by

Figure 7.1 Mature green coconuts.

Figure 7.2 Coconuts in various stages of maturity on a tree.

the Philippines, with Thailand, Malaysia, Indonesia, Sri Lanka, and Vietnam show-ing significant export growth in the last few years (Fig. 7.3). Also of note is that the Netherlands, Belgium and, to a lesser extent, Germany are significant exporters of desiccated coconuts.

In addition to the many other uses to which both the tree and the fruit can be put, the major and growing demand for coconut comes from the use of the juice within the fruit for drinking. Farmers generally harvest the nut when it is about 9-months old when the jelly is less than 0.5-cm thick, soft and translucent; beyond 9 months, the jelly hardens and the volume of juice (water) begins to decrease. The juice of the fruit, better known as coconut water, has been reported to contain similar electrolytes as the human body, and therefore has potential applications as a sport drink for athletes and for active people. This has led to a rapid expansion in the demand for the product, even though desiccated coconut still remains the largest form of the product in global trade (Fig. 7.3). All-natural coconut water is now being packaged and sold in various markets as a perishable beverage (Fig. 7.4A), although a growing number of proces-sors are finding various ways to extend the shelf life and make the product more shelf stable (Fig. 7.4B).

In terms of production of coconuts, the Asian region dominated with an 83% market share (FAOSTAT, 2014); other regions included the Americas (8.7%), Oceania (4.8%), and Africa (3.4%). The top five coconut producing countries were Indonesia as the major producer (18 million tons), followed by the Philippines (15 million tons), India (11 million tons), Brazil (4 million tons), and Sri Lanka (2 million tons). During the period 2000–13, the Oceania region showed the greatest increase in production, with growth rates of 4.6%; this was followed by Asia at 1.5%, the Americas by 1.1%, and

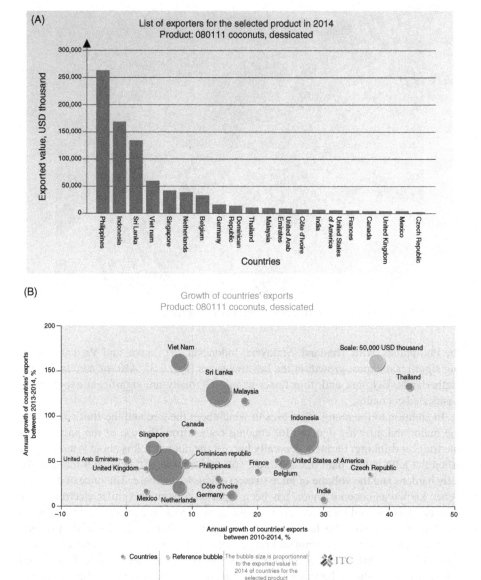

Figure 7.3 2014 Trade Data on desiccated coconut exports globally. (A) Exporters.
(B) Growth Rates.

Africa by 0.7% (FAOSTAT, 2014). Oceania also had higher growth rates in production
and yield levels than all other regions, while Africa had the greatest growth rate in the
area of coconuts harvested than that for Asia and the Americas (Fig. 7.5A–B).

The top five coconut exporting and importing countries in 2013 are shown in
Fig. 7.6A–B, respectively. In 2013, Indonesia, Vietnam, India, Malaysia, and Thailand

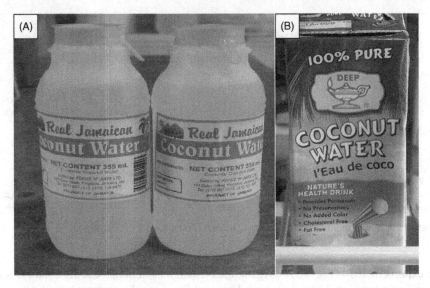

Figure 7.4 Packaged coconut water products marketed in developing countries.

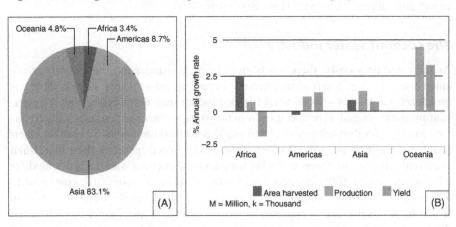

Figure 7.5 Coconut production share (A) and growth rate by region (B).
(*Source:* FAOSTAT, 2014. Crop production statistics. The Statistics Division of the Food and Agriculture Organization of the United Nations, Rome. Available from: http://faostat3.fao.org/home/E)

were the major coconut exporting countries in the world, with exports ranging from 50,000 to 230,000 tons. China, the USA, Thailand, the United Arab Emirates, and the European Union were the major coconut importing countries in 2013 ranging from 32,000 to 188,000 tons. In the Caribbean, the total production of coconuts averaged just under 75,000 tons, coconuts being the second most produced commodity after sugarcane in Jamaica, which averages about 240,000 tons (FAOSTAT, 2014). In terms of value of exports, countries from the far east remain among the largest exporters with Indonesia, Thailand, India, and Vietnam accounting for over US $200 million

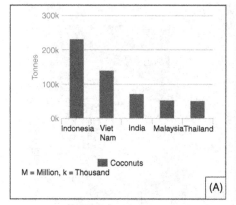 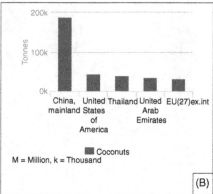

Figure 7.6 Top 5 exporting (A) and importing (B) countries for coconuts.

in exports in 2014 (Fig. 7.7A). Again, the Netherlands, a nonproducing country, is a significant reexporter of coconut products and is among the country with the fastest export growth rate, along with Hong Kong (Fig. 7.7B).

The coconut water industry

As indicated previously, there has been growing consumer interest globally in coconut water both as a refreshing, healthy beverage and as a sports drink. This has broadened the market opportunities for the product such that it has received significant investments and advertising from major stars in the entertainment industry, as well as major food companies in global markets with commitment to expanding their health and wellness portfolio (Euromonitor, 2011). Soft drinks data show that Brazil is currently the world's largest market for packaged coconut water; it accounted for 67% of the global 100% coconut water retail volume sales in 2010, compared to 47% in 2005, and 21% in 2003 (Euromonitor, 2011). Coconut water's success in Brazil indicates that major opportunities exist in other tropical countries where the drink is part of local beverage consumption culture and where virtually all coconut water in such markets is still consumed fresh, rather than packaged (Euromonitor, 2011). However, as these countries' packaged food and beverage markets continue to develop in sophistication, consumers will eventually be compelled by the advantages of purchasing their favorite liquid refreshment in an easily portable, convenient, and hygienic format to suit consumption occasions in more formal environments (Euromonitor, 2011). The United States is another major growth market, primarily because its consumer base comprises millions of ethnic Latin Americans who grew up on coconut water before emigrating to the United States. In Europe, coconut water has also started making noteworthy inroads in Germany and France (Euromonitor, 2011).

These growing market opportunities are driving the need for coconut water to be accessible in a more convenient format and have thus led to the development of

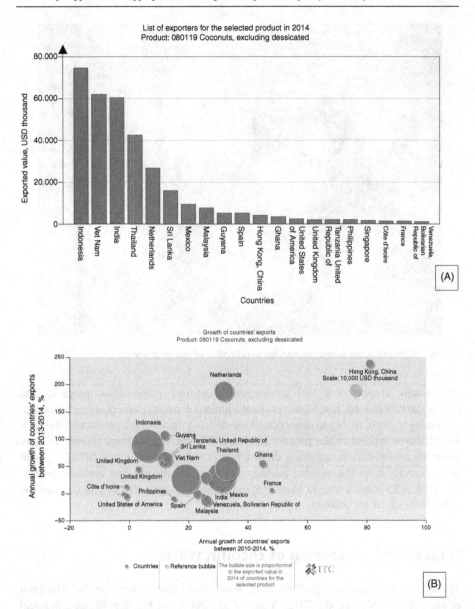

Figure 7.7 2014 Trade Data on exports of nondesiccated coconut products. (A) Exporters and (B) growth rates.

technologies for the preservation and sale of the product in a packaged format, including bottling. Bottling of coconut water not only reduces the cost of bulk transportation of immature coconuts over long distances, it also enhances the shelf-life, adds value, and generates income and employment for small farmers and coconut water processors (Rolle, 2007). The commercial production of coconut water has

Figure 7.8 Small-scale semicommercial production of coconut water.

traditionally employed a high-temperature/short-time preservation process. The product of this process has, however, found limited consumer acceptance in coconut-producing regions, owing to distortion of the delicate flavor of the product by the high temperatures applied in the process (Rolle, 2007). Most traditional producers have therefore reverted to the adapting the original process of cutting and filling coconut water directly from the nut into packages to a semicommercial or commercial setting (Fig. 7.8) or have sought to employ newer technologies that do not damage the delicate flavor of the product.

Quality and processing of coconut water

The chemical composition of coconut water has been widely studied (Jackson et al., 2004; Prades et al., 2012; Yong et al., 2009), as has the physicochemical properties of the product under various conditions of processing and storage (Awua et al., 2011). Compositionally, coconut water is made of about 96% water and 6% solids (Table 7.1). The main components of the solids in coconut water are soluble sugars, proteins, and minerals, which are made of a very relatively high level of potassium and, in contrast, low sodium. The composition of coconut water is highly dependent on the coconut variety and the stage of maturity of the nut (Jackson et al., 2004). For example, sugars range in concentration from 3.42% to 5%, proteins from 0.12% to 0.52%, fat from 0.07% to 0.15% and minerals from 0.47% to 0.87%

Table 7.1 Composition of coconut water from nuts at 6 and 12 months maturity

Components	6 Months maturity	12 Months maturity
Water	94.18	94.45
Total solids	5.82	5.55
Protein	0.12	0.52
Fat	0.07	0.15
Ash	0.87	0.47
Carbohydrates	4.76	4.41
Dietary fiber	ND	ND
Total sugars	5.23	3.42
Sucrose	0.06	0.51
Glucose	2.61	1.48
Fructose	2.65	1.43
Potassium	203.70	251.52
Calcium	27.35	31.64
Sodium	1.75	16.10
Phosphorus	4.66	12.79
Magnesium	6.40	9.44

Source: Yong, J.W., Ge, L., Ng, Y.F., Tan, S.N., 2009. The chemical composition and biological properties of coconut (*Cocos nucifera* L.) water. Molecules 14 (12), 5144–5164.

(Prades et al., 2012). Due to this favorable mixture of available nutrients, coconut water is very susceptible to spoilage which is a major consideration in its production and handling. There are a number of pre- and postharvest factors, as well as storage conditions that can negatively impact the quality of coconut water. Preharvest contamination by pesticide residues during production of the coconut, as well as heavy metals through soil or water contamination are major factors, both of which could cause illness (Rolle, 2007), although reports of illness due to coconut water are rare (Taylor et al., 1993) and none have been attributed to chemical intoxication. More important postharvest considerations include primarily contamination by microorganisms that could enter coconut water as a result of improper postharvest handling and processing techniques (Rolle, 2007).

Most coconuts are grown commercially by small landowners and the harvesting of coconuts occurs on these commercial farms by either climbing the tree using a rope or with the assistance of a power-operated ladder. Most commercial operations use dwarf varieties of coconuts that grow up to 15–25 ft. in height, while tall coconut trees grow to the height of 98 ft. (Pradeepkumar et al., 2008). Coconuts take 9 months to a year to fully ripen with coconuts growing together in a bunch tending to ripen at about the same time. If the fruit is to be harvested for the coconut water, it should be harvested at around 6–7 months after emergence (Pradeepkumar et al., 2008). Along with the timing, color is also an indicator of ripeness. Mature coconuts are yellow, brownish green, or brown, while immature fruit is bright green or yellow, depending on the variety.

Figure 7.9 Schematic of coconut water processing.

Coconut water processing

The steps in the processing of coconut water (Fig. 7.9) involve a number of activities from harvesting to bottling that must be carried out in a manner that ensures the production of high quality, natural coconut water for consumers. As the coconut matures, the amount of coconut water is reduced as the meat hardens. Fruits being harvested for water should be picked at the appropriate stage (Jackson, 2002; Rolle, 2007), the stages of maturity and their effect on the fruit being discussed by Jackson et al. (2004), Monro et al. (1985), Pue et al. (1992) and reviewed more recently in detail by Prades et al. (2012). If the tree is tall, a pole pruner may be of assistance during harvesting or using a ladder; if the tree is small or has bent from the weight of the nuts, it may be possible to reach the fruits easily and clip them from the palm using sharp pruning shears (Pradeepkumar et al., 2008). In any case, fruits can be harvested by either handpicking them, picking them using selected tools or picking them and lowering them to the ground (Pradeepkumar et al., 2008; Rolle, 2007). Coconuts should, however, never be allowed to fall to the ground during harvesting as this could lead to mechanical injury and facilitate the entry of microorganisms which cause spoilage of the coconut water within the coconut. Research conducted by Jackson (2002) show that coconut water collected from coconuts which had cracked when dropped from a height of 8 m, was cloudy (low % transmittance) as opposed to being clear in appearance and had a low pH, both of

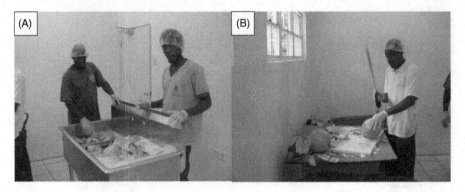

Figure 7.10 (A) Washing coconuts and (B) sanitary cutting of coconuts.

which are indications of spoilage of the product. Similarly, coconut water collected from dropped and cracked coconuts had a higher free fatty acid (FFA) content, than that collected from coconuts that were dropped and intact and those that were handpicked (Jackson, 2002).

Every step during the processing of coconut water is critical. The harvester, those who load, unload, and transport coconuts, those who cut coconuts, those who bottle and those who sell coconut water, are responsible for applying good practice to assure the quality and shelf-life of the final product (FAO, 2007). Only coconuts of the correct developmental stage (9 months), which are in sound condition and free from cracks should be used as sources of coconut water for bottling (Rolle, 2007); if they do not meet certain minimum standards, then they should be rejected. Rolle (2007) recommends that coconuts should be washed in potable water (Fig. 7.10A), soil removed to reduce the risk of transferring contamination during the collection of coconut water, followed by sanitation of the washed coconuts, which involves soaking the coconuts in a dilute bleach solution for about 15 min to further reduce the number of microorganisms on the surface of the coconut. This is critical as the water is removed from the fruit by cutting it open (Fig. 7.10B) and any contaminants on the surface of the coconut will likely end up in the coconut water if not removed prior to cutting. The coconut water is further processed using a filtration step that can be applied at different scales depending on the level of operation. For small-scale processors (like MSMEs[a]), the practice that should be applied first involves using a coarse filter, decanting the coconut water into a sanitized container equipped with a strainer lined with sanitized silk screen cloth or cheese cloth. The filtered coconut water should then be transferred to a cooling tank and cooled to 4°C (FAO, 2007) before being bottled.

FAO introduced a significant innovation to the industry when it received a patent from the United Kingdom in 2000 for the production of shelf-stable coconut water using membrane separation technology. This technology uses a cold processing treatment to reduce the microbial load of the product while retaining the typical flavor of coconut water. However, the high cost of the technology made it prohibitive for most small-scale processors who have continued to use less

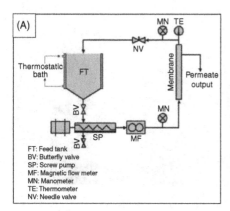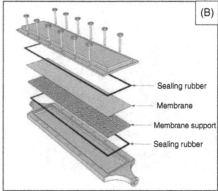

Figure 7.11 Schematic of a pilot microfiltration pilot unit (A) showing cross section of the membrane cells (B).
Source: do Nascimento Debien, I.C., de Moraes Santos Gomes, M.T., Ongaratto, R.S., Viotto, L.A., 2013. Ultrafiltration performance of PVDF, PES, and cellulose membranes for the treatment of coconut water (Cocos nucifera L.). Food Sci. Technol. 33 (4), 676–684.

expensive, appropriate processing technology that maintains the chemical composition, quality, and safety of coconut water. A number of larger companies, however, have employed the FAO patented microfiltration process in the making of coconut water. This is particularly so as further research has addressed the on-line differentiation of coconut water of varying maturities (Hahn, 2012) as well as some of the perceived flavor and stability deficiencies of product (Mahnot et al., 2014). The process involves a two-step filtration where the coconut water is prefiltered through a 0.80-μm membrane prior to final microfiltration with a 0.20-μm membrane (Fig. 7.11). The filtered coconut water is transferred to a cooling storage tank and then packaged using a volumetric filing machine. It is then stored until sold to consumers.

The storage of coconut water is also a critical step in the process as microorganisms can cause spoilage and affect the quality of the final product. Bacteria and yeasts are the predominant microorganisms associated with freshly bottled coconut water (Wizzard et al., 2002). These microorganisms multiply at a rapid rate at elevated temperatures and contribute to spoilage of the product. It is, therefore, critical that the temperature of bottled coconut water be maintained between 0 and 4°C, and that it is kept away from light during transportation and storage to assure its keeping quality and to enhance its shelf life (Jackson, 2002; Rolle, 2007; Wizzard et al., 2002). These considerations, as well as the practices used in harvesting, handling, and processing of coconut water were among the areas of focus during a series of studies which sought to characterize the best way to produce coconut water using a traditional approach. These studies and the important technology outcome for the industry are documented in "Case Study #1." An intervention in a set of processing facilities to improve the way they produced coconut water and its impact on the industry in a developing country are described in "Case Study #2."

Case study 1: development of objective easy-to-use approaches to manage the quality, safety, and shelf life of coconut water

Background

As was mentioned previously, the Food and Agriculture Organization of the United Nations (FAO) received a patent in the United Kingdom in the mid-2000s for their work in the production of self-stable coconut water using membrane separation technology. This technology was intended to use a cold processing treatment to reduce the microbial load of the product while retaining the typical flavor of coconut water. However, the high cost of this technology made it prohibitive for most small processors to adopt. In addition, for two medium-sized processors in different countries with which the authors were involved, the microfiltration also resulted in a removal of significant flavor compounds from the product, leaving a product that may have been safe, with a good shelf life, but which was bland and did not taste very much like coconut water.

Coconut water in Jamaica is cold processed using a simple manual filtration technique that was based on the same principle, but which resulted in significantly reduced shelf life. While this approach to processing retains much of the original flavor which otherwise would have been lost if a heat treatment was used, it results in a storage life of only about 1 week. At the time of this study, the coconut water industry had no objective method that it could use to assess the quality and potential shelf life of the coconut water being produced, significantly increasing the risk of economic losses should a particular batch not be of acceptable quality. The industry at the time was using sensory measures (visual, smell, and taste) to assess the quality and wholesomeness of coconut water. The primary way this was done was by the water quality of each individual nut being checked for its wholesomeness by sniffing the contents of the nut on the collection of coconut water for bottling. Apart from being subjective, this method increased the risk of contaminating the product prior to bottling, which could potentially result in reduced quality of the entire batch. Consequently, the authors' research team set out to develop an objective method through a project "Development of an Objective Method for Assessing the Quality of Coconut Water" that was funded by the Food and Agriculture Organization (FAO). The project involved a number of studies including physicochemical composition, enzymatic and microbiological content, aroma and lipid oxidation profiles of the nut and bottled water. This case study examines the approach to selected studies and the development of a method that could be used in commercial processing to ensure better quality and a longer shelf life of the refrigerated product.

The first phase of the assignment involved a study on the changes in composition of the coconut water during maturation. The second phase of research in the project involved the comparison of nut water from tall and green coconut varieties, the bottled product from four companies approved for sale by the local authorities,[b] as well as changes in composition, other indices, the microbiological quality of nuts, and bottled coconut water over time. The role of enzymatic activity in spoilage of the coconut water was also investigated.

Approach to the studies and methodology

For this study, coconuts (*C. nucifera* L) at an appropriate stage of maturation for water collection (Jackson et al., 2004) were obtained from the Coconut Industry Board of Jamaica (CIBJ). Coconuts were divided into two batches: one batch was used the other batch was used directly, not were washed by dipping in a 200 ppm hypochlorite solution, while the other batch (controls) were used directly without washing. All cutting operations were performed with a stainless steel knife, sanitized between cuttings with the use of a 100 ppm hypochlorite solution.

Sampling

Immediately after cutting, coconut water was taken directly from the nut, using a sterile pipette. Samples were also taken after decantation from the nut. The odor of each coconut water sample was noted upon sampling. For assessment of commercially produced coconut water, freshly bottled coconut water was purchased directly from a local processor and stored under commercial temperature conditions prior to all analyses. The whole nut, water, and shell weight of the coconuts were measured from representative samples and the percentage yield of coconut water calculated.

Fresh and processed product physicochemical composition

The chemical composition of all coconut water samples taken was determined on a daily basis over a period of 2 weeks. Water, total and soluble solids, reducing, nonreducing, and total sugars, ash, lipids, proteins, pH, total acidity (citric acid), and turbidity were determined in triplicate according to AOAC methods.

Enzymatic activity

Polyphenol oxidase (PPO) and peroxidase (PO) activities were determined in triplicate on all coconut water samples using the methods of Ponting and Joslyn (1948) and Fehrmann and Diamond (1967), respectively. The optimum pH (3.5–8.5) and temperature (5–60°C) for activity of the enzymes was investigated using the method of Campos et al. (1996). Enzyme activity under low temperature storage conditions (similar as a supermarket refrigerated cooler system) over a 2-week period were assessed to monitor its effect on product degradation.

Microbiological profile

Microbiological assessments were conducted on all coconut water samples over a 2-week period. Each sample was evaluated in duplicate. Total viable organism and spore counts (mesophilic and psychrotrophic), as well as specific organisms including coliforms, *Eschericha coli*, *Staphylococcus* spp., Clostridia, *Listeria* spp., *Bacillus*, and yeasts were assessed. Mold growth was also determined. Microorganisms were further characterized, where relevant, using biochemical tests, including catalase and lecithinase production, lipase activity, starch hydrolysis, production of acid from arabinose, fructose, glucose, lactose, galactose, sucrose, maltose, and sorbitol (Harrigan and McCance, 1976; Speck, 1992).

Rancidity

The extent of lipid oxidation and hydrolysis of all samples were assessed by determining the peroxide value (PV) and FFA content, respectively according to AOAC with slight modification and expressed as milliequivalent (meq.) peroxide per kilogram coconut water.

Other parameters

Changes in composition and other physical parameters including titratable acidity, pH, Brix, and turbidity were also measured.

Physicochemical properties

Volume and percentage water, total and soluble solids, pH, total titratable acid (% citric acid), ash, lipid, protein, and turbidity were determined according to AOAC methods (1980). The percentage water was determined by taking the weight of the whole nut, water, and shell of each coconut. Soluble solids were determined using an Abbe refractometer at 25°C and expressed as °Brix. The pH of the samples was measured using a digital pH meter (Accumet). Total titratable acid was determined by titrating 10 mL of each sample with 0.1 M NaOH solution, using phenolphthalein as the indicator. Calcination of charred coconut water at 550°C in a muffle furnace was used to determine the ash content. Dried samples of the coconut were assessed for crude fat content by solvent extraction using a Soxhlet apparatus. Turbidity of the samples was determined using a spectrophotometer at 610 mm, relative to distilled water. All analyses were carried out in triplicates.

Physicochemical composition of Jamaica coconut water during maturation

This study reports the physicochemical composition of coconut water from dwarf trees measured at 7, 8, 9, and 10 months during its maturation. The volume of water present in the whole nut increased on average from 233 to 504 mL from maturity stages 7 to stage 9. This is known to be due to the fact that as the nuts mature they become larger and thus their water holding capacity increased. A marginal decrease was observed between stages 9 and 10, which could be attributed to hardening of the nut jelly at stage 10, resulting in a reduction in the volume of water present in the nut (Jackson, 2002). The crude fat content in the samples increased as the coconuts became more mature. Research by the Caribbean Food and Nutrition Institute has suggested that it is the constituents of the nut water that is used to form this "jelly," and so the closer the coconut gets to the stage of "jelly" formation, the higher the fat content of the water. The protein content showed a relatively large increase from 0.03% to 0.10% from 7 to 10 months of maturation. The water also became more turbid with maturity, a finding that could possibly be related to the increased ion concentration that was also observed. The ash content of the coconut water, which represents the overall mineral content, did not show any significant difference between the stages.

It can be concluded, therefore, that the mineral content of coconut water remains relatively constant during stages 7 to 10 of maturation.

The total solids are a measure of the dry matter that remained after the moisture was removed from the samples, and is reported to be due to the increased concentrations of ions such as phosphate, sulfate, chloride, and fluoride as coconuts mature (Pue et al., 1992). The pH of the samples at different stages showed slight variation during maturation; overall, a marginal decrease was observed as the coconuts matured from stages 7 to 9, with a larger decrease from stage 9 to 10 (Jackson et al., 2004). This trend was the opposite of that observed by Pue et al. (1992) and may have been as a result of the difference in the coconut varieties or the locations studied. It has been shown that coconut composition varies depending on the geographical and climatic factors at different locations (Neto et al., 1993). The percentage titratable acid (%TA) for the coconuts confirmed increasing acidity with maturity as the more mature samples were more acidic. There was a significant increase in the citric acid content from stages 7 to 9.

In summary then, the most significant finding from this aspect of the study was the change observed in the volume of nut water, which increased during development from 233 to 504 mL, with the greatest quantity at 9 months. Fat, protein, soluble solids, acidity, and turbidity also increased steadily with maturity; while pH and ash showed variation throughout maturation.

While the carbohydrate profile was not assayed in the study on stages of maturity, as expected, the coconut water samples increased in sweetness with maturity. Another aspect of the study examined the impact of storage temperature and time on the carbohydrate profile of the product. This involved studying the changes in total sugars (%TS), reducing sugars (%RS) and nonreducing sugars (%NRS) over a period of 7 days (Fig. 7.12). The data show that while there was some (marginal) decline in sugar content for the product stored at 4°C for 7 days, the overall changes were not significant. Conversely, storage of the coconut water at 26°C led to very significant declines in all categories of sugars by day 7 (Fig. 7.12). This suggested that the sugars present in coconut water were being transformed during the period of storage, a finding that will be examined in more detail in subsequent sections.

Microbiological mediation and enzymatic activity during maturation and spoilage of coconut water

The study examined the presence of microorganisms in whole nuts and nuts cracked during picking and falling to the ground. As has been previously indicated, Jackson (2002) reported that the likely entry of microorganisms into coconuts which had been cracked during picking and falling to the ground were a likely cause of the greater spoilage associated with coconut water made from these nuts. This informed the advice provided by the FAO in the industry document outlining best practices for harvesting and handling coconuts to be used for coconut water (Rolle, 2007), which recommends that coconuts should not be allowed to fall to the ground during harvesting. This study also documented the role of selected microorganisms in the spoilage of coconut water during storage at 4, 18, and 26°C for a period of 5 days (Figs. 7.13 and 7.14).

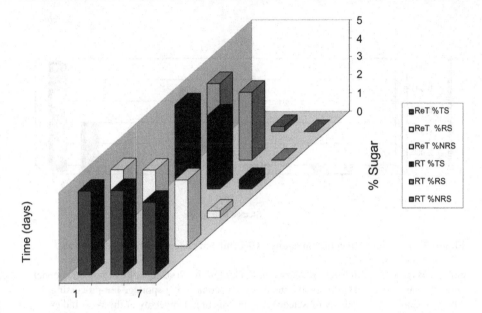

Figure 7.12 Changes in carbohydrate profile of coconut water over 1, 3, and 7 days at 4 and 26°C. *ReT*, Refrigerated temperature—4°C; *RT*, room temperature (26°C); %TS, % total sugars; %RS, % reducing sugars; %NRS, % nonreducing sugars.
Source: Jackson, J., 2002. Coconut Water Quality Evaluation. Final Report submitted to the Food and Agriculture Organization of the United Nations (FAO), Rome.

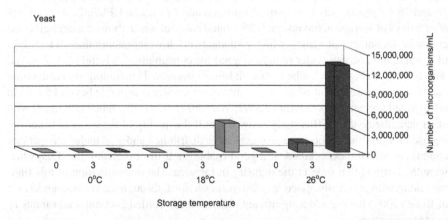

Figure 7.13 Changes in yeasts counts (CFU/mL) of coconut water over time at different temperatures.
Source: Wizzard, G., McCook, K., Jackson, J., Gordon, A., 2002. Quality of bottled coconut water during storage; pH as a reliable indicator of coconut water spoilage pre-processing. Abstract, Caribbean Academy of Science Annual Meeting, University of the West Indies, Mona, Jamaica.

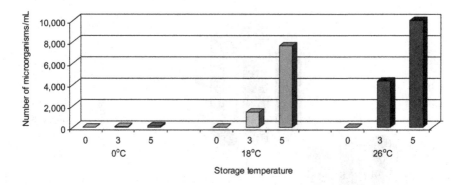

Figure 7.14 Changes in coliform counts (CFU/mL) of coconut water over time at different temperatures.
Source: Wizzard, G., McCook, K., Jackson, J., Gordon, A., 2002. Quality of bottled coconut water during storage; pH as a reliable indicator of coconut water spoilage pre-processing. Abstract, Caribbean Academy of Science Annual Meeting, University of the West Indies, Mona, Jamaica.

Lactic acid bacteria (LAB) have been shown to be active in the spoilage of coconut water over time (Wizzard et al., 2002). The findings of this study (Fig. 7.13) indicated that yeasts, along with known spoilage agents such as LAB, were major causes of the rapid deterioration of the product, when unrefrigerated. At the higher temperatures of 18 and 26°C, yeasts increased to over 3 million and 12 million CFU/mL, respectively after 5 days of storage. Conversely, refrigerated coconut water showed a negligible increase in the number of yeasts, which remained very low, indicating that up to 5 days of storage, the coconut water remained good for consumption, a trend also observed with coliforms (Fig. 7.14), albeit at much lower numbers. This finding was also borne out by sensory assessment which found the coconut water acceptable beyond 5 days at refrigerated temperatures (Jackson, 2002; Wizzard et al., 2002). When combined with other data, the results of this study showed that if the quality of the coconut water being used for bottling is good, the product can be distributed and sold under refrigerated conditions with the expectation of a good usable shelf life exceeding the 3 days that were the norm at that time in the industry in Jamaica. The dissemination of this finding, along with others discussed in subsequent sections, through the recommendations of Rolle (2007) have made a significant impact on the bottled coconut water industry in many countries.

Enzymatic activity

The study evaluated the activity of important enzymes during the process of maturation and also during spoilage of coconut water over time, some findings from which have been reported (Jackson, 2002; Jackson et al., 2004; Wizzard et al., 2002). The activity of enzymes is heavily influenced by the pH of the environment, all enzymes having optimal pH ranges. Due to this, the study also evaluated the pH of the samples in which the enzyme activity was being monitored. The initial pH of the samples was 5.07, which reflected fairly low acidity. After 1 week, there was a significant reduction in pH particularly in the samples stored at 18 and 26°C to about 3.74 and 4.32, respectively. There was also a simultaneous increase in the titratable acidity (%TA). These changes were likely due to the increased microbiological activity observed during storage, particularly of LAB and yeasts (Fig. 7.13), as well as chemical reactions (lipolysis) that would release acid as a by-product.

PPO and peroxidase (PO) are two commercially impactful and important enzymes found in coconut water, and are involved in catalyzing the oxidation of phenolic compounds, resulting in pink to brown discolorations. The activity of both enzymes were found to decrease over time in the samples that were stored at 18 and 26°C This was not observed in samples stored at 0°C for PO activity. These findings were to be expected, as the pH of the samples stored at 18 and 26°C was outside the optimum pH range for PPO (pH = 6.0) and PO (pH = 5.5) (Wrolstad et al., 2000). Increases in turbidity of the coconut water samples was also observed during storage. This was more pronounced in samples stored at 18 and 26°C, particularly after 3–5 days, and coincided with the increase in numbers of microorganisms in the samples during that same period (Figs. 7.13 and 7.14). PPO and PO activity caused by endogenous and microbiological enzymatic activity, which results in a discoloration as well as acid production, could precipitate coconut proteins, and also lead to increased turbidity. Conversely, there was no significant change in turbidity of the samples stored at 0°C, primarily because of the reduced chemical and microbial activity in those samples.

Acid value (AV) and %FFA, indirect measures of lipase activity (hydrolytic rancidity), were found initially to be 0.97 and 0.34, respectively. Both AV and %FFA showed a significant increase during storage particularly, at 18 and 26°C, suggesting increased lipase activity at the higher temperatures. Lipase hydrolyses fats in the coconut water to form fatty acids. Some short chain fatty acids that are produced during lipolysis may result in offensive odors and likely accounted for the rancid aroma, and also the reduction in pH and increased %TA observed. The lipase activity observed may be due to endogenous and microbiological sources. PV is a measure of oxidative degradation, possibly due to lipoxygenase (LOX) activity (Wrolstad et al., 2000). LOX catalyzes the oxidation of polyunsaturated fatty acids to form hydroperoxides, the initial product of lipid oxidation. The hydroperoxides are converted to aldehydes, ketones, and other compounds (Wrolstad et al., 2000), which are responsible for the rancid off-flavors and aromas observed. The values obtained showed that as the time of storage and temperature increased, the PV of the samples increased, indicating that increased lipid oxidation, and therefore greater detectable and objectionable oxidative rancidity were also taking place. This was confirmed by the sensory monitoring done and demonstrated the role of enzymatic degradation in the spoilage of unrefrigerated or temperature abused, natural coconut water.

Correlations between dependent and independent variables of coconut water quality, first reported by Wizzard et al. (2002), are shown in Table 7.2. They support the findings of the study that endogenous and microbiological enzymatic activity

Table 7.2 **Correlations between the independent and dependent variables of coconut water quality**

Dependent variables			
Independent variables	%TA	Turbidity	pH
PPO	−0.758**	0.507*	0.868**
PO	−0.236	0.338	0.278
PV	0.466	−0.506*	−0.533*
FFA	0.976**	−0.867**	−0.972**
Temp	−0.652**	0.577*	0.646**
Time	0.471*	−0.555*	−0.526*
APC	0.460	0.595**	0.449
Yeasts	0.643	−0.602	−0.779*
LAB	0.493*	−0.629**	−0.481*

*Significant $p < 0.05$.
**Highly significant $p < 0.001$.
Source: Wizzard, G., McCook, K., Jackson, J., Gordon, A., 2002. Quality of bottled coconut water during storage; pH as a reliable indicator of coconut water spoilage pre-processing. Abstract, Caribbean Academy of Science Annual Meeting, University of the West Indies, Mona, Jamaica.

contribute most to the spoilage of coconut water. Specifically, they demonstrated that the %TA and pH of the water were influenced by the FFA content and that %TA and pH were also influenced by the activity of PPO ($p < 0.001$). The data also suggested turbidity was most greatly affected ($p < 0.001$) by the bacterial activity (LAB and APC) and the FFA content of the coconut water ($p < 0.001$).

Conclusions

These studies showed that the loss of quality and shelf life in coconut water appeared to be due primarily to endogenous and microbiological enzymatic activity that produced rancid off-flavors, odors, and a reduced product pH. The storage life of coconut water was found to be highly dependent on temperature, and could be significantly extended by storage at temperatures below 4°C. This is critical for retailers since wide temperature variations in refrigerated storage areas could adversely affect quality. The studies also showed that high initial microbial load in the coconut water samples resulting from poor harvesting, handling, and processing practices were likely to significantly reduce the commercially viable shelf life of the product. Since coconut water was (and remains) cold processed in many commercial operations, there is no kill step to reduce the microbial load. Extreme care must therefore be taken during harvest and postharvest handling to reduce the initial and final microbial load that ends up in the commercial product. Also, while the transparent bottles and lighted storage conditions for coconut water in supermarkets may be desirable for consumers, these are likely to accelerate the oxidative degradation and rancidity of the product leading to a shorter shelf life.

Based on the findings reported, it was felt that the use of an objective index to eliminate the blending of compromised, poor quality coconut water in with good quality water as the water from different nuts are aggregated prior to storage or bottling would significantly reduce the incidents of large-scale spoilage in commercial processing. The data presented showed and subsequent testing proved that pH measurement done on the water from individual nuts prior to adding/blending it in with other coconut water already collected was an effective way of identifying and eliminating poor quality coconut water that was likely to spoil in short order. It was therefore recommended, and has been employed in processing facilities that a simple, calibrated pH meter is used to check the pH as an indicator of water quality since the primary factors contributing most to spoilage correlate quite well with pH (Table 7.2). This would reduce problems of subjectivity in nut assessment prior to processing. Narrow range pH paper should be used to reduce cross contamination if adequate cleaning of the pH meter electrode in between sampling cannot be assured. The optimal pH range is 5.0–5.4. Anything below this indicates potential or likely spoilage.

It was also recommended that coconuts should be harvested by handpicking to reduce any cracks that would allow entry of microorganisms. In addition, a postharvest dip of at least 50 ppm (5%) sodium hypochlorite should be used to reduce the microbial load on the surface of the nuts. The chlorine strength of the dip solution should also be monitored throughout processing to ensure adequate bactericidal activity. Finally, alternative packaging and storage conditions at retail should be explored to seek to mitigate the effect of commercial distribution conditions on the deterioration of quality in refrigerated coconut water.

Case study 2: modification of the production of bottled coconut water to improve its quality, safety, and shelf life

Introduction

The coconut water industry today is a US $300+ million industry (Fig. 7.3) and continues to grow rapidly. Companies now are introducing more sophisticated packaging and, as discussed in the previous case study, looking at various technologies to retain the natural taste of coconut water and extend the shelf life of the natural product. Manufacturers are using a variety of approaches to capitalize on the popularity of the beverage, including adding coconut water to a variety of other juices to make blends, several of which are gaining popularity in North America and Europe. The more traditional buyers want to be able to get pure natural coconut water (or as close to it as possible) and other coconut by-products such as coconut milk[c] in a convenient format as and when they want it. These products are therefore now becoming a staple in discerning stores such as Walmart in the United States (Fig. 7.15). In the early 2000s, however, the industry was just emerging from the doldrums of a period in which coconut oil and associated coconut products were reported as being very bad for health because of the widespread use of some hydrogenated coconut oils with the attendant issues of high *trans* fats and saturated fats. At that time, those producing coconut

Figure 7.15 Coconut water and coconut milk products on the shelf in Walmart.
Source: André Gordon, 2016.

water in developing countries did not have the information or technology available today and faced the issue of trying to market a very delicate, sensitive product with a very short shelf life. In this context, an intervention to increase the viability of the industry across the Caribbean region and implement best practices and its impact may be instructive as a case study.

The nascent coconut water industry in the Caribbean

The level of sophistication in the coconut water industry varied across the world in the early to mid-2000s, depending on the country in which the product was being made. In Brazil, for long a world leader in coconut technologies, coconut water was already a staple in many cities throughout the country while the level of sophistication varied in the countries of the Far East. In the Caribbean, Jamaica, Dominica, St. Lucia, and Trinidad and Tobago had well-developed coconut industries, mainly focusing on coconut oil produced from copra.[d] Of these countries, only Jamaica had a formal industry producing and, in some cases, exporting coconut water, all other Caribbean countries not yet having transitioned from coconut oil as a focus of the industry to coconut water. This project involved assessing the current state of the industry in Jamaica and St. Vincent and the Grenadines and developing solutions for the challenges being faced to improve the viability of the industry.

Approach to the interventions

To prescribe useful solutions to the coconut water industry, it was necessary to get an understanding of exactly how the industry was structured, where their markets were, how the product was produced, and what the technical problems were. This required a detailed assessment of the industries in both countries, including a diagnostic of the challenges that were being faced, primary among which was a pink color which arose from time to time and a very short shelf life (3 days or less) for the refrigerated product. This was followed by the development of a specifically tailored approach for each company to address the issues that arose as a result of the assessment, training, where applicable for their staff and guidance with the initial implementation of food safety and quality systems. The information presented in this case study is an amalgamation of what was found in different producer firms across the industry in the markets examined.

Industry practices

An assessment of the practices across the industry showed that, with minor differences, most firms approached the harvesting and handling coconuts in a similar manner, as shown in the Fig. 7.16. Coconuts were typically grown on large privately owned estates of land holdings and picked by itinerant pickers who then transport them to the bottling plants who receive the nuts and prepare them for recovery of the coconut water. The receival and handling practices (Fig. 7.17) varied between plants and, along with harvesting, were the areas in which there were opportunities for improvement. The bottling process used by most producers included cutting open the nuts (Fig. 7.18A), recovering the water (Figs. 7.18B and 7.19), then either putting this into storage in varying types of holding tanks (refrigerated and nonrefrigerated) prior to filling (Fig. 7.20), capping (Fig. 7.21), and refrigeration and/or freezing of the bottled product. The bottles being used at the time were high density polyethylene (HDPE) or PET (in the case of St. Vincent and the Grenadines, only PET was used at the time), and closure (capping) was done under ambient atmosphere (i.e., no modified atmosphere treatment, e.g., nitrogen flushing, was used).

The companies involved in the bottling of coconut water in the Caribbean in 2003/2004 were all what would be classified as *small and medium sized enterprises* (SMEs). It was to be expected, therefore, that there would be some practices that would need improvement. The areas that needed improvement were from the harvesting of the coconuts through to the delivery to customers. These were indicated in the detailed reports to each of the entity and are summarized in the overall approach recommended to the industry which is presented in Fig. 7.22 and detailed in subsequent sections.

Deficiencies in existing practices

The harvesting of the coconuts as was being done at the time was not in accordance with best practices, subsequently reported in Jackson et al. (2004) and Rolle (2007). The practice which predominantly allowed the coconuts to fall to the ground, was one which would result in increased contamination of both the surface of the coconuts and, as indicated by Jackson (2002), a significant increase in spoilage due to contamination of the water because of the damage to the nut. The receival and handling of the nuts also predisposed the coconuts to further contamination as these were often transported in uncovered vehicles with workers walking on the nuts which were placed directly on the floor of the vehicles (Fig 7.17A–B). The nuts were then received at the plant and typically stored directly on the floor until ready for processing

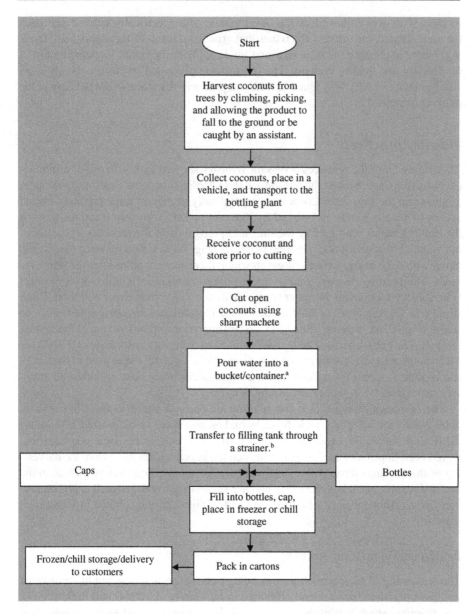

Figure 7.16 The coconut water production process in the Caribbean 2003/2004. [a]Some bottlers poured the water through a strainer to remove extraneous matter at this step. [b]If not strained earlier.

Figure 7.17 Delivery of coconuts to bottling plant (A) and storage (B).

Figure 7.18 Cutting and coconut water harvesting (recovery) at two plants.

Figure 7.19 Straining coconut water through muslin into holding container.

Figure 7.20 Filling coconut water into bottles.

Figure 7.21 Capping (closure) of bottled coconut water at an SME.

(Fig. 7.17B). The attire of the staff, in general was not in keeping with best practice with staff wearing their own clothes, sometimes covered by an apron (Figs. 7.18–7.21), other times without. This is a breach of GMP and was also a practice that could contribute to contamination of the product. The nuts were taken from storage (on the floor) and cut to pour the water into containers, prior to bottling. In many cases, there was no washing or sanitizing of nuts prior to cutting. The physical surroundings and housekeeping in the areas where cutting was done was also deficient in some cases (Figs. 7.18B and 7.19). The cutting of the nuts was done with machetes that were cleaned at varying frequencies, often insufficiently, and in some cases, they were not sanitized after washing. The coconut water was strained through muslin or a metal strainer to remove any particulate in facilities but the frequency of changing the cloth or muslin, or of washing and sanitizing the strainers was not standardized in most cases, was not documented and the effectiveness of cleaning was not verified. In most firms, filling was done under fairly sanitary conditions (Fig. 17.20B) but in

Figure 7.22 Coconuts being held in trolley during storage and handling in a plant.

a few cases, the practice of filling was such as to possibly put the product at risk of contamination (Fig. 7.20A). The capping of the bottled product was done manually, with homemade implements such a wooden "hammers" being used to seal the cap on in one case (Fig. 7.21). Refrigerated or frozen storage was generally good except that there was little or no records kept for traceability or of the temperatures of the freezer/chiller.

In summary then there were many areas throughout the process of preparing bottled coconut water that contributed to the quality (and potential food safety) and shelf life issues being experienced by manufacturers. Many of these could be addressed by implementing GMPs and, in some cases, having the firms develop and implement an HACCP program. The major concerns and quality/shelf life/safety implication of major observations are shown in Table 7.3. Based on these observations, recommendations as to how to mitigate the concerns and deliver good quality, safe coconut water with a longer shelf life than was currently the case at the time are also presented (Table 7.3). Although there was a recommendation that coconuts were to be stored elevated and kept from contacting the ground (Rolle, 2007), one important recommendation not included in any previous report on the handling of the fruit is the use of trolleys to receive and store the product during the precutting stages (Fig. 7.22). Another important recommendation was the measurement of pH. A summary of the recommended quality assurance practices to improve the quality, safety, and shelf life of coconut water is presented in subsequent sections. The revised process (Fig. 7.23) recommended to SME bottlers shows the steps to be taken in the handling and bottling of high quality, safe coconut water with a good refrigerated shelf life.

Table 7.3 Observations on practices at SME coconut water bottlers and recommendations to improve quality, safety and extend shelf life

Selected observations	Implication	Recommended solution
Coconuts allowed to touch the ground or fall during harvesting	Contamination of the water inside the nut with spoilage or disease-causing microorganisms	Nuts should be picked and lowered by rope to a waiting trolley/container
Transportation of coconuts to the factory on the ground in open vehicles; staff feet in contact with nuts	Contamination of the exterior of the nut with spoilage or disease-causing microorganisms	Nuts should be transported in covered vehicles and elevated off the floor by pallets or slip sheets; staff should not walk on nuts
Nuts delivered directly onto the floor of the bottling plant	Contamination of the exterior of the nut with spoilage or disease-causing microorganisms	Nuts should be delivered into a container that is elevated off the floor or placed on pallets
Staff were improperly attired with street clothes not covered	Potential contamination of the product by improperly attired staff	Implement a strict garment policy based on best practice
Nuts were cut without being cleaned	Nonremoval of potential contaminants from the exterior of the nuts prior to cutting	Nuts should be washed with a suitable food-grade detergent and a brush, if necessary,[a] and the exterior surface sanitized prior to cutting
Housekeeping in the cutting area was poor with cut coconuts being left on the floor, becoming an attractant for flies	Potential to cause insanitary conditions in the cutting area that could result in contamination of the product	Implement strict good housekeeping and GMPs
Implements used for cutting nuts were not sanitized with sufficient frequency to provide assurance of cleanliness at all times	Unclean cutters or machetes could become a source of microbial contamination for the coconut water	Establish a validated frequency of cleaning and effective sanitation of machetes and other cutting tools
Filtration was done through cheesecloth, muslin, or screens but no validated frequency of changing or cleaning these existed	Unintended contamination of the coconut water from contaminated or recontaminated screens or straining cloths	A validated frequency must be established, documented, implemented, and monitored (with appropriate records) for the changing and sanitation of screens, cheesecloth, and muslin cloths used
Bottles and caps were cleaned by washing but there was no indication of the effectiveness of the cleaning and sanitation	Potential contamination of the product with physical matter and microbial contaminants from unclean bottles or caps	Bottles and caps should be washed with potable water containing at least 5 ppm residual hypochlorite[b]
In some instances, product was held at ambient temperature while awaiting filling	Potential for initiation of microbial growth during the holding period, resulting in shorter shelf life	Store all product at 4°C or below while awaiting and after bottling

[a]See recommendations in Rolle (2007) or in FAO (2007).
[b]Based on the recommendations of McGlynn (2015).

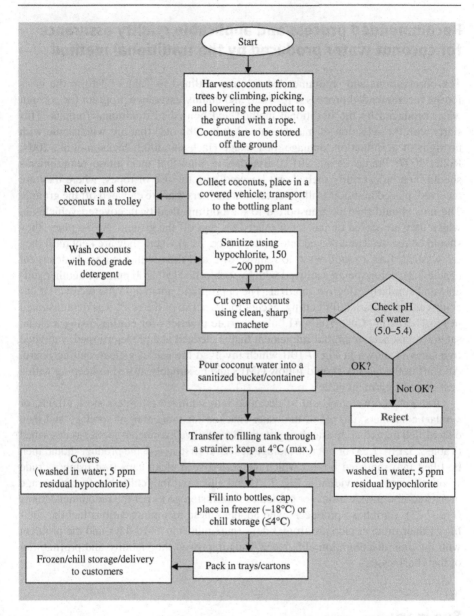

Figure 7.23 The recommended process for coconut water bottling by SMEs.

Recommended process and applicable quality assurance for coconut water produced by the traditional method

The observations and recommended solution outlined in Table 7.3 form the basis for the recommended process (Fig. 7.23) and quality assurance program for coconut water produced by the traditional method and bottled in traditional formats. This starts with the selection of mature nuts (7–9 months old) that are wholesome with no signs of germination, damage, or leakage (Jackson, 2002; Jackson et al., 2004; Rolle, 2007; Prades et al., 2012). Bunches of nuts that meet these requirements should then be carefully picked and lowered with the use of ropes, where they are received, placed in trailer trolleys or other containers that are elevated off the ground. The nuts should then be transported to the bottling facility in covered vehicles in which they are stored on pallets or otherwise kept off the ground. At the plant, they should be received and placed into trolleys (Fig. 7.22) where they are kept until they are washed, first a rough wash to remove dirt, and then with food grade detergent (Table 7.3), before being sanitized with hypochlorite (150–200 ppm) or another suitable and similarly effective sanitizer. The appropriate practice in the receival of coconuts and preparation for cutting, except for the attire of the staff, was demonstrated in Fig. 7.10A in "Case Study #1." Appropriate practices for cutting, using a clean, sanitized machete or similar implement that is cleaned at a predetermined, validated frequency is shown in Fig. 7.10B which involved the use of a clean cutting board. All staff in the facility should follow GMPs and be suitably attired in keeping with a best-practice garment policy.[e]

Once cut, the water should be decanted into sanitized (stainless steel, HDPE, or similar) containers, the pH of the water checked for indication of spoilage and then either filled immediately after being strained to remove extraneous matter as described (Table 7.3; Fig. 7.23) or put into storage at 4°C. If first stored, the product should then be filled into sanitized bottles which are sealed with clean caps (not using a crude wooden hammer as evident in Fig. 7.21) and either put into chilled or frozen storage or packed into trays/cartons for chilled or frozen storage as per the recommendations (Fig. 7.23). Caribbean producers who adopted these recommendations had the shelf life of their product increase from 3 days (maximum) to 7–10 days and the problem with the pink discoloration eliminated. This improved the viability and profitability of their businesses.

Summary

Two case studies were presented that examined different but related aspects of the coconut water industry. The first set of studies examined fresh-from-the-nut and bottled coconut water, including that stored at 4, 18, and 26°C over a 1-week period to determine the factors that contributed most to spoilage and the role of various constituents and practices in determining quality and shelf life. The possibility that the pH could be used as a quick and reliable indicator of nut spoilage was specifically explored. The second case study explored the practices within the industry at the time (2003/04) with a view to improving safety, quality, and shelf life through the application of improved practices and technology.

In the first set of studies, the impact of the microbial flora, enzymatic activity (lipase, PPO, and peroxidase), and oxidative stability was determined on bottled coconut water over its shelf life. Within 1 week of storage, a rancid odor was evident in bottles stored at 18 and 26°C, while those stored at 4°C maintained much of the aroma of the original bottled coconut water. Aerobic and lactic acid producing bacteria, yeast, and mesophilic spores were found at significant levels all samples assessed at the end of the 10-day period. PPO and peroxidise (PO) activity decreased overall during storage. Hydrolytic (lipase activity) and oxidative rancidity (as measured by PV) increased significantly within 3 days of storage at the elevated temperatures of 18 and 26°C. There was also a significant decrease in Brix and pH, and an increase in titratable acidity and turbidity which coincided with the degradative processes occurring during storage. The studies showed that oxidative and hydrolytic degradation both of which resulted in lower pH, contributed significantly to organoleptic spoilage of coconut water. They indicated that the determination of pH, which is simple, economical and rapid, is capable of giving a reliable estimate of the spoilage status of coconut water prior to processing, thereby mitigating the potential use of water of compromised quality and resulting in longer refrigerated shelf life and better quality coconut water.

The second case study showed that many of the commercial practices being employed, from harvesting through to packaging, ignored the science behind the production of high quality, more extended shelf life of refrigerated coconut water. It showed how simple changes and the application of GMP and basic scientific principles could result in at least a doubling of the shelf life of refrigerated coconut water from 2–3 days to 7–10 days and reduction in the incidences of discoloration due to PPO (Jackson, 2002). The recommendations included better sanitary handling of the fruit and the plant environment, as well as including the measurement of pH as a quality control measure to reduce the contamination of the product and deliver higher quality coconut water.

Both of these case studies and the information derived from them have found application in the rapidly growing coconut water industry globally. The body of work reported here and selected findings and recommendations formed the basis for the definitive industry manual developed by the FAO (Rolle, 2007) and used around the world for the industrial production of coconut water. If the process outlined in Fig. 7.23, the measures summarized in Table 7.3 and the attendant recommendations are followed by any processor, the production of high quality, safe coconut water with a refrigerated shelf life of 7–10 days by the traditional method is assured.

Endnotes

[a]MSMEs—micro, small, and medium-sized enterprises.
[b]The Jamaica Bureau of Standards was the regulatory authority of import in approving the process and facilities.

[c]Coconut milk is the liquid expressed when the hardened coconut jelly (or coconut meat as it is sometimes called) is grated or ground and squeezed to release a milky sweet white liquid, rich in energy, and selected minerals. It is popularly used in cuisine in the Caribbean, Southeast Asia, and South America.

[d]Copra is the dried coconut meat or kernel from which the oil is extracted. It is derived from the coconut jelly which, as it ages, becomes the coconut meat as it thickens and hardens.

[e]The garment policy defines how staff throughout the facility should be attired in each of the operational areas and how garments are to be handled, in general. This includes footwear, head wear, and other protective clothing.

Case study: FSQS in solving market access prohibition for a vegetable product—callaloo (*Amaranthus sp.*)

8

A. Gordon
Technological Solutions Limited, Kingston, Jamaica

Introduction

Amaranthus viridis is among the plants grown in developing countries that is slowly finding its way into the cuisine of developed countries. It is found in Vietnam, where it grows in the wild (Tanaka and Nguyen, 2007) and is part of local diet, and parts of Africa (Fondio and Grubben, 2004), where it is native as well. It is used in the traditional cuisine of southern and north-eastern India, where it is known by various names, including *kuppacheera* and *cheng-kruk*, respectively. In the Maldives, it has been a part of the diet for centuries (Romero-Frias, 2003), and it is also among the greens used in the European country of Greece, where it is called *vlita*. Along with *Amaranthus spinosus* L., which it closely resembles, it is known as *tanduliya* in Sanskrit in various parts of India, where is also used a medicinal herb in traditional Ayurvedic medicine (Nair, 2004). While the leaves and stem are the major part of the plant used in the cuisines mentioned previously, the seeds of *A. viridis* and other *Amaranthus* sp. are the main parts of the plant that have been used in the traditional cuisine of Mexico and other Central American countries for centuries. In these countries, *Amaranthus* seeds form an important part of the diet, particularly as a source of protein and other minerals (Gordon, 2015a*).

A. viridis (Fig. 8.1) and *Amaranthus dubius* are both closely related and difficult to differentiate, and are also known in the Caribbean as callaloo,[a] a very popular vegetable in Jamaica and those areas of North America and the United Kingdom where large Jamaican or Caribbean populations are present. It is also increasing in popularity, as these Caribbean nationals have influenced their African–American friends to use the product instead of spinach (*Spinacia oleracea*) or collard greens[b] (*Brassica oleracea*), particularly in the southern states of the United States when both of these vegetables are unavailable during the winter months. The product is exported from Jamaica to the Caribbean region and North America, where it has become a major menu item particularly during the winter and spring months. Callaloo comes from the Amaranthaceae family and is a highly nutritious product, which, like collard greens, is rich in vitamins A and K (Fig. 8.2) and is a good source of potassium and calcium (Gordon, 2015a), particularly for the poor and marginalized. This staple represents a new trend among the more ethnically diverse populations in the major metropoles, who are eating increasingly greater amounts of the foods that they were used to in their home countries (Hallam et al., 2004).

Figure 8.1 Callaloo (*A. viridis*).
Source: Adapted from Gordon (2015a).

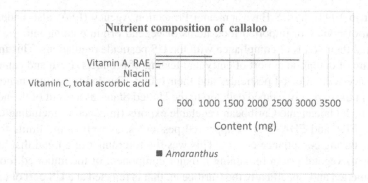

Figure 8.2 Vitamin content of callaloo (*Amaranthus* sp.). Vitamin A is presented in international units (IU); *RAE* - Retinol activity equivalents.
Source: USDA National Nutrient Database 2014 (Adapted from Gordon, 2015a*).

Due to its increasing popularity and, consequently, an increasing demand for the product, manufacturing firms from Jamaica, the main Caribbean country in which callaloo is eaten in this form, have been commercially producing and exporting the product to North America and the United Kingdom for decades. This has further been fuelled by the growth in the popularity of Jamaican and Caribbean cuisine in the last decade or so and the development and expansion of quick-serve restaurants and other dining establishments that make the cuisine available commercially in major cities, such as London, Toronto, New York, and Washington, DC, among others. Consequently, canned callaloo had, over the years, become the second largest product by volume made and exported by the micro-, small-, and medium-enterprise (MSME) food-processing sector in Jamaica for its own label or for major exporting/trading entities. The trade was so important by the early 1990s, that callaloo exports were approximately 18% of all canned food exports from Jamaica by value, and a critical product for the health of MSME agroprocessing export sector.

The growth in the trade continued uninterrupted until the early 1990s when issues of compliance with regulatory requirements, including the presence of pesticide residues and foreign matter arose in both the United Kingdom and the United States, the latter being more definitive about its insistence on compliance at the time. In mid-1993 to early 1994, the US Food and Drug Administration (FDA), in assessing imported items of vegetal (plant) origin found a high level of imports being noncompliant with its defect action level (DAL) standards for foreign matter in foods.[c] This included all plant-based food imports from Latin America and the Caribbean, a major source of the US' horticultural imports, and also extended to canned vegetable items, such as canned callaloo. As a result of this, items found to be violative[d] were put on automatic detention or detention without physical examination (DWPE)[e] as it is called in the US regulatory language. This meant that all canned callaloo imports from Jamaica were automatically prohibited entry into the US market unless it could show that it was compliant with the required standards. The specific problems that arose with noncompliance with the FDA-stipulated limits for foreign matter, their classification, and how compliance was assured are addressed in the first case study.

Later in 1994, the US Environmental Protection Agency (EPA) also undertook a program of review of imported food items of vegetal origin coming into the United States and their status of compliance with the US pesticide regulations. This included a program of detention at port of entry and testing of products (fresh and canned) for compliance with allowed pesticides and their limits. This led to another major interruption in the export of canned callaloo to the United States as a result of the noncompliance of Jamaican and Caribbean vegetable exports (in general), including callaloo with the EPA and FDA's list of approved pesticides, as well as the limits found in products during compliance checks. This was the forerunner of a trend that has seen products of vegetal origin remaining a major component of the imported food from Latin America and, specifically the Caribbean, that is rejected at a US port of entry for pesticide and contamination issues (Gordon, 2015a). To address this matter, a careful science-based approach in assuring the safety and compliance with the regulations had to be implemented. This issue with the use of pesticides on callaloo and similar vegetable products to be exported from developed to developing countries is addressed in the second case study in this chapter. Both case studies (the first and second) should be useful to exporters of foods of vegetal origin as they seek to access the US market. This is even more particularly so as the Current Good Manufacturing Practice and Hazard Analysis and Risk–based Preventive Control for Human Foods rule (21 CFR 117), also called the Preventive Controls Final Rule, are fully enforced, beginning September 2016 (Food and Drug Administration, 2015a).

Case study: production of callaloo with compliant extraneous matter levels

Background to case

Agricultural products play a crucial role in Jamaican exports and is tied to the Jamaican agroindustrial sector, which harvests, processes, and packages several traditional foods for the local market and, critically, for the export market. Callaloo, sometimes called spinach or *bhagi* in other parts of the Caribbean and which is indigenous to Jamaica, is a major component of the Jamaica diet. The plant produces dark, large green leaves, which are usually boiled or steamed and consumed as a vegetable. Over the years, several farmers across the country have been involved in the growing of callaloo specifically for export, including supplying the processors who pack for the export market as well. The vegetable is traditionally grown on farms with plot sizes ranging from approximately 0.25 ha for the small farmers to 2–3 ha for larger farmers. During this process, a range of pesticides is often used to protect the crop from pest damage. It is the use of these pesticides and the insect pests being targeted that engaged the industry with the US regulatory authorities in 1993–94.

Currently, pesticides are among the major reasons for rejection of products of vegetable origin at the ports of entry in the United States (Saltsman and Gordon, 2015; ECLAC, 2011) At the time of this case study in the early 1990s in the US market, there were also a significant number of pesticide-related issues that were arising as a result of federal and state regulatory activities. As a result of this, at that time (in 1993), the FDA had announced an expanded plan for pesticide monitoring, while the EPA had initiated special reviews of allowed pesticides with a view of reassessing the "adequacy" of their allowed limits and, by extension, the risk to humans from consuming them. They required brand owners to provide proof of the efficacy of pesticides on specific pesticide/crop combinations, postharvest intervals[f] (PHI), as well as the safety of the level of residual pesticides typically found remaining on the crop. Several major pesticide manufacturers consequently elected to drop quite a few crops from the label of their products rather than support the expensive field trials required for their approved use on those crops, with the risk of failing in the field trials and the attendant financial cost being major concerns. Malathion, a widely used pesticide in both developed and developing countries at the time, was an example of one such pesticide where the registrant had decided to remove several crops from its label.

In this context, a developing country farmer who was using (and continue to use) pesticides made in developed country markets that had not specifically been tested for or approved for use on "indigenous" crops faced the challenge of not being able to prove safe usage of the pesticide on a specific crop. This required that either the manufacturer or the user undertake the field trials required, find another way to prove the safety of the pesticide/crop use combination, or, failing this, discontinue the use of the pesticide and find a suitable alternative that was approved and registered for use on the crop. For developing country farmers and exporters, this was far more of a challenge than it may appear, as many had been using various pesticides imported from third-world country markets for many years, relying on the fact that these were also used in the home markets as a proxy for their safe use on their crops. In many cases, the pesticide would have been initially approved for use on the same group of corps or a crop similar in nature to those approved by the EPA or its sister regulatory authorities in Canada, the European Union, Australia/New Zealand, and other similar markets. In several cases, however, significant usage drift would have taken place and the crops on which the pesticides were being used were removed from even the family of corps they were originally approved for [e.g., a pesticide approved for cabbage (*Brassica* spp.) being used for roots and tubers].

Against this background, Jamaican farmers planting callaloo (*Amaranthus* sp.) for export, processors manufacturing canned callaloo, and large export distribution firms faced a challenge in 1993 when the matter of the use of specific pesticides to control insect pests on callaloo, as well as the insect burden of the imported product itself, fell under FDA scrutiny. As EPA has an arrangement with the FDA to act on its behalf and enforce pesticide residue limits compliance at port of entry, the FDA was the agency with which the matter had to be addressed. This case study describes how the problem of foreign (extraneous) matter (XM) in callaloo exported to the United States arose, was diagnosed, and a solution developed and successfully implemented. It includes a discussion of the complication for the use of unregistered pesticides and how a resolution to this was arrived at, considering the need to also have an effective pest management to meet US limits for foreign matter, while meeting pesticide residue limits as well.

Callaloo production

Callaloo, a major product for several SME manufacturing plants in Jamaica, is processed by several factories for a number of different brands, with its production in many processing facilities being the second largest sustained volume, underlining its importance and the economic value. Its production is quite labor intensive from farm to canned product, and includes land preparation, chemical treatment (fertilization and pesticide application), irrigation, and reaping (Fig. 8.3), followed by processing if canned or packaged products are to be produced (Fig. 8.4). The processing involves first stripping and cleaning the product to remove the hard, woody areas of the stem

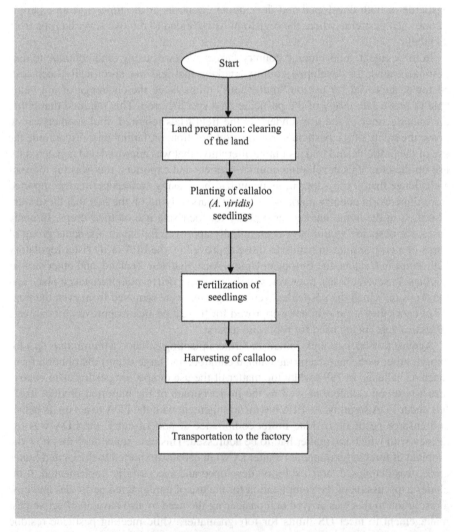

Figure 8.3 Process flow for agricultural management (land preparation to harvesting) for callaloo (A. viridis).

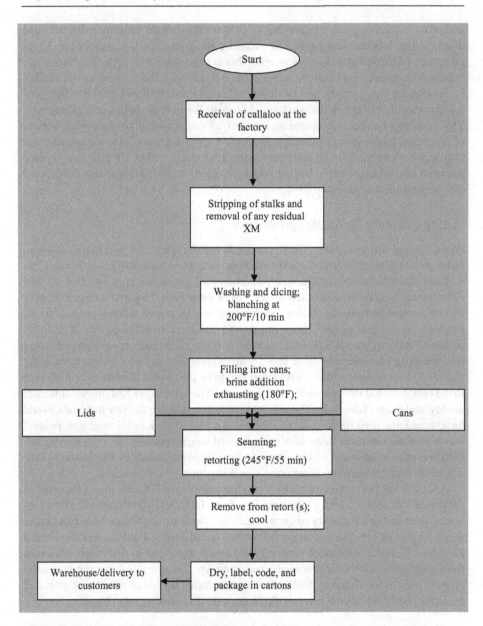

Figure 8.4 Process flow for production of canned callaloo (A. viridis). XM, Extraneous (foreign) material.

and any visible XM, followed by washing and dicing/particle size reduction (often done manually). The vegetable is then blanched, filled into cans, brine added, the vegetable exhausted with steam to remove any gases trapped within the plant tissues, the can lid affixed, and then seamed to form a hermetic seal (Fig. 8.4). The callaloo is then processed in a retort using a preestablished scheduled developed by a process authority; cooled (typically in cooling trough or tanks) under chlorinated water; and dried, coded, labeled, and packed in cartons prior to storage in a warehouse. Cost-effective production that delivers product that is compliant with all requirements demands strict monitoring of all the stages of growth, reaping, and processing of callaloo.

Among the major production facilities is Central Food Packers Limited (CFPL) situated in Central Village on the outskirts of Spanish Town, in the parish of St. Catherine. This company has been in operation for over 25 years, and produces a wide variety of processed foods. It began callaloo processing approximately 20 years ago and was also approved by the FDA in 2004 for having a satisfactory HACCP program in place to export canned ackee to the United States [see Gordon (2015a) regarding the process that was required to facilitate this].

Quality assurance for callaloo production

At the time of this case study, CFPL produced callaloo under its own brand, and also did contract packing for other large distribution companies including Lasco Foods Limited and GraceKennedy and Company Limited (GK), both based in Jamaica, as well as importers located in the United States. All of these firms required stringent quality control to meet both the regulatory requirements and their own internal product specifications. At that time as well, the oversight and management of the quality of all foods produced and exported by GK fell under the purview of the author, who was therefore also involved whenever any issue arose with exports. For callaloo, before processing was started, there were certain checks that had to be performed to ensure conformance to the established standards on the raw materials. These included XM, microbiological quality, and color (Table 8.1). These would determine whether the raw materials would be accepted and used for processing. The texture, leaf-to-stalk ratio, stem size, percent leaves (drained), percent stems (drained), drained weight, percent salt, commercial sterility, vacuum, headspace, and can seam indices were measured for the finished product, while time/temperature monitoring was done during the processing.

A significant percentage of CFPL's production was for GK, the major exporter of the product from the country. As is evident from Table 8.1, significant effort was invested in ensuring the quality of canned callaloo made for the Grace brand and other major brands in Jamaica and exported to the United States, Canada, and the United Kingdom. Samples were taken randomly and assessed for quality and safety attributes from each lot or batch after processing was complete. Despite this, however, it was the product packed for Grace by CFPL that was the initial source of the import alert and therefore the firm played a pivotal role in the case discussed in this chapter.

The issuing of the import alert

Callaloo was being exported to the United States from Jamaica for many years, without problems. However, in May 1993, XM tests were conducted by the FDA on a range of callaloo brands being exported to that market. Insect fragments in excess of the allowable DALs for spinach and also for leafy greens were found to be present in the callaloo samples. This raised concern from the FDA that imposed a strict testing

Table 8.1 Quality assurance (QA) characteristics for canning of callaloo in 1993

QA characteristics	Value/limit(s)	Comments
Raw Materials		
XM	Absent	Visual assessment at plant
Color	Dark green	Visual assessment at plant
Finished Product		
Microbiological quality		
Commercial sterility at 35°C	Sterile	Sometimes done at 55°C
Sodium chloride salt (%)	1.0–1.5%	—
Vacuum (mm Hg)	127	—
Headspace (mm)	5–8	—
Texture	Soft, but not too soft	—
Drained weight (oz.)	15	—
Can seam analysis	Various measurements	Canadian standards
Leaf to stalk ratio (70/30)	—	—
Stems < 7 mm	—	—
Leaves 60–70% when drained	—	—
Stems 30–40% drained	—	—
XM	Absent	BAM[a] method
In Process		
Time–temperature monitoring	245°F/55 min	—

[a]The FDA's Bacteriological Analytical Manual method.

regime on all subsequent shipment coming from Jamaica to the United States. The FDA's prohibition was announced with the advisory excerpted here.

Specimen Charge

Form FDA 2678a (6/86)

  " The article was adulterated when introduced into and while in interstate commerce, and is adulterated while held for sale after shipment in interstate commerce within the meaning of said act 21 U.S.C. 342 (a) (3) in that it consists in part of a filthy substance by reason of the presence therein of insects; and of a decomposed substance by reason of presence therein of decomposed spinach."

This was an excerpt from the document issued by the Office of Enforcement, Division of Compliance Policy to GK at the time of the detention. The wording used was standard and so the references to *filthy substance* and *decomposed spinach* were due to the standard wording typically used in correspondence when a violative imported shipment was found rather than the actual presence of "filth" or "decomposition" in the product. It simply meant that, in accordance with the published FDA standards, the DAL's for XM (called filth) and vegetable matter that could be described as aged (called decomposed) was found on examination. The FDA then issued a guidance document to all exporters from Jamaica to advise them of the standards required and the steps they were expected to take if they intended to reenter the market with product that would not be found to be violative. The guidance issued to processors and exporters was as follows.

FOOD AND DRUG ADMINISTRATION COMPLIANCE POLICY GUIDES 7114.24.

Dated December 08, 1988

Chapter 14 - Vegetables

Subject: Spinach, canned or frozen - Adulteration involving insects

REGULATORY ACTION GUIDANCE

The following represents the criteria for direct reference seizure to the Division of Compliance Management and Operations (HFC-210) and for direct citation by the district offices.

1. Aphids, Thrips or Mites

Aphids, Thrips or Mites singly or in combination, average 50 or more per 100 grams.

2. Spinach Worms (Caterpillars)

Two or more measuring 3 mm or longer (Larvae or fragment) whose aggregate length exceeds 12 mm, are present in 24 pounds.

3. Leaf Miners – Leaf Miners of any size average 8 or more per 100 grams OR Leaf Miners 3 mm or longer, average 4 or more per 100 grams.

Remarks:

Seizure involving these products must be discussed with the U S Department of Agriculture. Submit the following information to the Division of Regulatory Guidance (HFF-310) and await reply before proceeding:

Sample Number   Date of Shipment
Article Involved   Dealer
Amount of Lot   Shipper
Codes   Analytical Conclusions

It should be noted that although the violative product was a low-acid canned food that was within the FDA's jurisdiction and because the source of this initial prohibition was pests, which fell under the purview of the US Department of Agriculture (USDA), the FDA collaborated with them in this process.

Approach to addressing the market access prohibition (import alert)

The imposition of the import alert on first, Grace-branded callaloo, and then all product coming from Jamaica led to the development of a comprehensive approach to address the FDA's concerns. A task force (TF) involving the author, other members of

GK's technical support team,[g] and staff from other areas of the company, as well as interested contracted farmers and contract packers was formed in June 1993, and mandated to solve the problem in the shortest possible time. In its assessment of the root cause of the problem, the TF discovered that not only were the pesticides being used at the time ineffective in controlling the common pests for the callaloo, but the manner of cleaning and washing the vegetable prior to processing was not sufficiently effective in removing any insects or insect parts remaining in the product. This meant that it would be impossible to ensure an ongoing compliance with the DAL for filth (XM) unless significant changes were made to the way the vegetable was grown (Fig. 8.3) and processed (Fig. 8.4). What ensued was a 13-month program that transformed the way the crop was grown, handled, and processed that ensured compliance with all of the import requirements of the United States.

The TF that was empanelled included quality assurance (QA) specialists, a crop scientist, an agronomist, a biometrician, an analytical chemist, manufacturing specialists, engineers, and persons with specialization in export trading. They were supported by the technical resources of the Ministry of Agriculture's extension arm and GK's own Grace Technology Centre (GTC), a fully equipped, food science and technology research and development, QA, and technical extension facility. The team developed a comprehensive approach to solving the problem that examined and addressed:

- Crop variety selection.
- Agronomic practices, including details of plant spacing, husbandry, and pest management.
- Postharvest handling and transportation, from farm to factory.
- Good agricultural practice (GAP).
- Preprocessing handling, particularly with a view to reduce the insect burden, should any be found on the crop.
- The in-plant production process.
- QA and product safety.

The approach also addressed the possibility that the growing operations by some farmers had, from time to time left detectable pesticide residues on the callaloo. With the renewed and stringent monitoring being undertaken by the FDA, any such occurrence would present another set of challenges and require ongoing shipment by shipment testing, a costly undertaking that was to be avoided.[h] The TF set about the planning and implementation of a series of targeted tasks, studies, research, in-field trials, and process modifications that were expected to result in a sustainable solution to the problem. The process was estimated to take 9 months to complete, but this was later extended to 13 months because of unforeseen and uncontrollable circumstances, particularly with the agronomic side of the project. Specifically, the TF had members address each of the following:

1. Verifying the allowable limit for XM in canned (or fresh) callaloo (*Amaranthus* sp.).
2. Identifying and/or developing the best integrated pest management (IPM)–based approach to be used as part of the crop husbandry program for callaloo. This included determining the options available for pesticide use during the growing cycle.
3. Determining what the optimal agronomic practices were, including plant spacing, planting, harvesting, and postharvest handling that would minimize the extent, if any of insect presence in the crop.

4. Transportation and receival and storage best practices.
5. Best practices in handling and preparing the in-coming raw material for thermal processing.
6. Optimal in-plant handling and processing practices.
7. Development of a comprehensive set of standards and a QA program for callaloo, including reviewing and modifying specifications (if necessary) for the growing, handling, and processing of the callaloo. This would include also a new set of QA criteria and control limits.

The process that ensued was one in which the TF team members worked simultaneously to address the information gaps in each of the critical areas required to attain full compliance with the US regulations. The field trials and agronomy was undertaken at the same time as efforts to improve the efficacy of the manufacturing process and the removal of XM was being pursued. Throughout the entire process, the technical members of the TF were gathering information and managing the technical components of the project, while remaining in constant contact with the regulators. This was important to ensure that these efforts would have a positive outcome. The details of each component of the solution development involved are outlined in the sections further.

Optimizing the agronomy of the growing of callaloo (*Amaranthus sp.*)

It took a period of exactly 12 months prior to June 1994, for the agronomic aspects of growing callaloo in a manner that would allow the final canned product to be compliant with the US regulations to be accomplished. The crop had a growing cycle of 6 weeks before harvest and further 6–8 weeks prior to replanting after the first harvest. It would therefore require a block of 3–4 months for each set of trials, once these were determined to be necessary, hence the extended timeline for completion. *This (the time required per growing cycle) is an important consideration for any firm or industry dealing with a market access prohibition involving agronomic practices.* This should be among the primary considerations when planning an approach to deal with problems, such as these, as realistic timelines that allow for the work required to be done thoroughly and to the satisfaction of the regulators will have to consider any petition for readmission of the product into their market.

For callaloo, after months of optimizing the process, a compliant system for callaloo production was developed and fully documented, specifying on a week-by-week basis the amount of fertilizer to use per acre, as well as the pesticides to be used. This was done with the help of a field officer who was employed to monitor the plots, keep record of the time the crop was planted, when it was transplanted, and when it was fertilized, including the type and amount of fertilizer used. The types/brand of insecticides and the concentrations, as well as the rate of application was noted. The officer had to ensure that the plants were sprayed on time and that the correct PHI was allowed before reaping. After reaping, the proper procedures were applied to ensure that the vegetable was taken to the factory in clean lorries that were free of contaminants. In this case, the use of pesticides containing *cypermethrin*, which was widely used in the United States on cabbage, tomatoes, and other fruits, was explored and its applicability for callaloo was determined. This was tied to the production practices in the factories and consideration of required yields and waste to advise amounts required and, hence, the pesticide usage. A conversion rate of the number of pounds of callaloo required per case of finished product (including wastage) was established and tied to pesticide usage. The wastage would largely be due to the need to remove the thick, high-cellulose stems from the plant, which would need to be trimmed, as well as any flowers that may be present. This formed the template for studies for other pesticides that were also investigated.

In addition to cypermethrin (branded Karate), other pesticides available in the Jamaican marketplace and known to be used in the United States were tested. Trials were also done with malathion, the pesticides branded Decis, Thuricide, Diazinon, Lannate, Belmark, Kocide, and Basudin. For all selected pesticides, a further set of trials was needed to determine the impact of plant population on pesticide efficacy. This would provide definitive information on the impact of plant population density on pesticide usage, and hence on how the efficacy of each potential pesticide/pesticide cocktail impacted both residual XM and pesticide residues on the crop, when harvested. Indicative information on some of the more efficacious pesticides tested during the series of trials is provided in Table 8.2.

Having completed the field trials with the crop and examining a range of variables and options over the 12-month period, it was concluded that callaloo treated with Lannate, Karate, and Basudin produced a significantly greater yield than callaloo treated with Malathion. These findings allowed the finalization of the process for producing callaloo that would be compliant with import requirements once the levels of both residual pesticides and XM in the product were acceptable. This was predicated on the assumption that the pesticide in question could be used on the callaloo, in keeping with USDA/FDA registered pesticides/crop combinations.

Table 8.2 Selected pesticides assessed for the integrated pest management (IPM) program for callaloo production

Pest and damage caused	Control[a] (oz./acre)	PHI (days)
Leaf roller: light green caterpillars that feed on the undersides of leaves, folds leaves, and feeds within the protected fold	Lannate: 8 Karate: 16 Belmark: 8	2–3 7 7
Spider mites: tiny red insect-like pests, which attack the leaf surfaces, leaving silver area on the underside, and yellow areas on the upper surfaces	Karate: 16	7
Cucumber beetles: chew on the leaves	Lannate: 8 Karate: 16 Belmark: 8	2–3 7 7
Leaf spots: white roughened areas on leaves	Kocide: 35	7

[a]Active ingredients were: Lannate: methomoyl; Karate: cypermethrin; Belmark: fenvalerate (a pyrethroid); Dupont Kocide 2000: copper hydroxide.
PHI, Postharvest interval.

The following were therefore required on completion of the field trials:

1. For callaloo, a definitive statement from the FDA as to whether there were FDA or other (specified) tolerances for pesticide residues and foreign matter for green leafy vegetables.
2. A definitive conclusion from analysis of experimental data as to which if any of the pesticides used gave predictable results regarding foreign matter content using tolerances for spinach if there were no other more generous tolerances.
3. The finalization of an agronomic protocol, including full documentation of all practices in a standard operating procedure (SOP) format and manual that could be provided to the farmers selected to guide them in the practices that would be acceptable. The manual would also include the data collection forms (records) required to accompany each shipment of the callaloo from farm to the factory.[i]
4. A comparison of the pesticide residues detected against tolerances allowed for spinach (*S. oleracea*) and the group leafy vegetables, a grouping that specifically included *Amaranthus* sp. (callaloo).

These would be addressed by the gathered data used to inform the QA program.

Development of quality assurance criteria

Handling the regulatory interface

In advance of and while the field work was being undertaken, the first series of tasks required the technical team involved with the project to develop an approach that would facilitate collaboration with the regulatory authorities in the United States (i.e., the USDA, EPA, and FDA). This would allow that the outcome of the efforts to address the problem would be acceptable to the importing market's regulatory bodies with certainty, a critical consideration when dealing with issues involving both sanitary and phytosanitary measures imposed by them and also any technical barriers to trade that may exist. In this particular case, the prohibition of exports as a result of insect matter in the crop would be an *SPS* issue, as would any issue arising from the pesticides used and residues present in the product, if any. The matter of the actual level of insects and insect parts allowable would be a *TBT* issue. While this may seem unimportant from a practical perspective, it determines how the issue is approached and which agency would be the one with the authority to agree satisfactory compliance measures.

Agreeing acceptable limits for callaloo

In this case, it was determined that the USDA would be required to approve the agronomic issues, as well as issues of classification of the plant, in collaboration with its sister agency, the FDA. Note that both the import alert and industry guidance document issued by the FDA (presented previously in this chapter) specifically mentioned spinach (*S. oleracea*), not callaloo (*Amaranthus* sp.), as the US authorities did not appear to have any standards specifically for callaloo. This necessitated a discussion and agreement with the US authorities on both DALs for *thrips*, *mites*, *aphids*, (*spinach*) *worms*, and *leaf miners*, as well as any other insects that may be a part of the pest pool for the crop. Another matter that also had to be addressed with the USDA and the FDA (on behalf of the EPA, at the time) was the specific pesticides allowed for usage on the callaloo, as well as the limits of any residues that may remain on the product. Again, all of the documentation available was for *spinach* but not *callaloo*. This is an important aspect of this case study, as this is likely to be a situation that most traditional crops from developing countries will face unless the manufacturers of the pesticides include the specific traditional crop in the trials when developing limits and the regulatory authorities of the importing country formally recognize them.

In this case,

1. The matter was handled by approaching the FDA Compliance Officer at the port of entry (Miami) where the DWPE was first imposed and requesting guidance as to what was required to make future shipments compliant. This led to discussions with CFSAN's team with whom ongoing dialogue was established as a way that could lead to mutually beneficial outcomes.
2. The FDA team agreed to provide information on the DALs for spinach and collard greens, either of which could hopefully be used for *Amaranthus* sp., as well as if the pesticides could be used on the crop.
3. The regulators agreed to assist with information regarding the allowed pesticides on crops entering their jurisdiction, particularly those similar to callaloo, which could be considered for approval.
4. The regulators agreed to share information and accept reasonable submissions with regard to both a DAL and pesticide residue limits (MRLs) for callaloo.

This approach and the agreements arrived at were critical, as without them the resolution of the issue would not have been possible.

Defect action levels

In November 1993, the technical team within the TF decided on an approach to gather data and the studies needed to fill any gaps in the data required to prepare a comprehensive case to petition the US regulators regarding the DALs and MRLs for pesticides on callaloo.[j] A series of experiments guided by the biometrician's expertise in the sampling of callaloo for the presence of insects (XM) and pesticide residue were agreed. The targeted XM level, the actual XM limit, the strategy being used to control XM levels, the field trials that it was subjected to, and the scale-up of experimental plots to commercial ones were also discussed. The unavailability of data on the efficiency with which each process stage (during production) reduces the XM load in the product was also discussed and a study to provide this data was commissioned from the GTC. It was decided that the biometrician would visit one of the lead growers' farm, as well as a lead contract packer, CFPL, during the production of callaloo to develop a statistically sound sampling schedule for the assessment of XM and pesticide residue for callaloo. The series of studies, including agronomic trials, were then commenced.

It had earlier been concluded from reference to 40 CFR[k] that callaloo (*A. viridis*) was, at that time, listed under ISO 34 (F) (9) (IV) (A). As a consequence, the technical team would need to establish the DALs for the leafy vegetables (except *Brassica*) group, that is, "extraneous nonvegetable matter" levels for *A. viridis*. The USDA was asked to advise what these levels were or refer the team to the pertinent publication where this information may be found. In response, it was advised that according to the FDA experts, the DAL for spinach would most likely apply to canned callaloo. They also indicated that they would send a booklet on all DALs currently used by the FDA. Later, in December 1993, in response to another request from GTC, the USDA forwarded information as to what pesticides may be applied to callaloo based on the other US-grown crops it was similar to.

Research had indicated that the tolerance set for Lannate in spinach (by the EPA) was set at 6.0 ppm. Consequently, any analytical checks being done for this pesticide had to be able to detect levels that were at least 5.0 ppm. Where no residues were detected at a 5.0 ppm level, it could safely be concluded that callaloo sprayed with Lannate could satisfy EPA requirements for spinach (if the recommended PHI was adhered to). When taken with the results reported for XM (insects and insect parts, see further), it meant that this pesticide was both detectable with the analytical technology available nationally at the time[1] and compliant for both pesticide levels and foreign matter. The technical team had also found that Karate, with the active ingredient cypermethrin, could also possibly be used on crops, such as callaloo, as permethrins (which are *pyrethroids*) were allowed for use as a spray for similar crops in the United States. Once it was determined that this active ingredient could also be determined at the local laboratory below the tolerance level set for it at the time (5 ppm), as also was the case for two other potentially useful pesticides, Decis (*deltamethrin*) and Diazinon, they could be used in trials.

Another important piece of information that came forward from research in March 1994 was that although the FDA (EPA) did not have any limits set for callaloo (*Amaranthus*) for pesticides and XM, it did for the classification of "leafy greens," a class of vegetables that already included "leafy amaranth." This meant that two potential approaches existed to settle the matter of both the pesticide residues and XM. The options appeared to exist to seek approval under the same terms as applied to spinach (which the FDA was already inclined to support, based on the import alert) or "leafy greens," which its classification system appeared to allow. This information proved useful and was very instructive toward the efforts of the industry to comply with existing EPA regulations and facilitated the experimental trials on callaloo, which were expected to be completed shortly, at that time.

Gathering the data

Agronomic practices

The data-gathering process for the agronomic information required was described previously. This provided information that allowed the research team to be confident that Lannate (with a 3-day PHI) could be satisfactorily reaped in 7 days, and that Depel and Thuricide would show no residues, as they affect biological, not chemical, control. Trials were also done using pyrethroids to determine their suitability for use. All crops were assessed at GTC, and the local laboratory used the pesticide residue analyses to

Table 8.3 **Effect of selected pesticides on residual XM in callaloo** (*Amaranthus* sp.)

Pesticide used	XM found (quantity/100 g)	XM found (particle size in mm)
Lannate	1.3	<3
Karate	2.3	<3
Basudin	2.6	<3
Malathion	4.1	<5

determine the efficacy of the treatments. The results of the test for XM on nine cans of callaloo selected according to the experimental design by the biometrician in the first three of six harvests produced are shown in Table 8.3.

While these were preliminary data, they were sufficiently encouraging to prompt more exacting, detailed studies designed by the biometrician, which supported the conclusion that with the proper practices, callaloo could be grown from the November to March period (the winter months in North America) and produce acceptable levels of XM. These practices would also deliver compliant, (nonviolative) pesticides residues using Lannate, Thuricide, and Depel alone or in rotation, as long as the growing was supervised.

Samples of canned callaloo produced under different agronomic management regimes and sampled as indicated by the biometrician were sent to Campden Chorleywood Food Research Association (CCFRA)[m] for pesticide residue analyses. This followed the completion of all of the trials and served as a further confirmation of the results obtained using the local laboratory. The results obtained are presented in Table 8.4.

The results indicated that although very low levels of residues were found for Decis, the residue levels for Belmark exceeded the allowed limit of 0.05 mg/kg (ppm) on leafy greens and spinach. This indicated that even with the improved agronomic practices, it would likely be difficult to be certain that none of the product being

Table 8.4 **Results of analysis of canned callaloo produced using the revised IPM program for callaloo production for selected pesticides residues**

Pesticide	Active ingredient	Detection (mg/kg)
Decis	Deltamethrin (pyrethroid)	0.01
		Not found
		Not found
Belmark	Fenvalerate (pyrethroid)	0.70
		0.19
		0.06
Lannate	Methomyl (carbamate)	Not found
		Not found
		Not found

shipped would be violative if Belmark were used in the IPM program. On the converse, Lannate was undetectable in the samples of canned callaloo tested. Another pesticide, Karate had also shown reasonable efficacy in sustainably reducing levels of XM found (Table 8.3) However, it was not as effective as Lannate and also had mixed results in terms of the recovery of pesticide residues from the canned product (data not shown). Consequently, the results of this study confirmed earlier findings and also confirmed to the technical team that Lannate should be the pesticide of choice around which other approved pesticide should be rotated in the IPM program for the production of callaloo.

Processing practices

While the field trials were ongoing, a project to develop and implement industry best practices was also initiated. A GK-owned factory, working with the technical team from the TF, undertook trials to determine whether the current, more traditional, and manual process for the preparation of callaloo could be improved. This involved the design, procurement, and testing of a new, semimechanized production line for the vegetable, inclusive of mechanized washing, and filling, prior to retorting. Several trials were done to optimize the operation of the line. When optimal efficiencies were achieved, trials on the ability of the line to reduce the insect burden and, hence the XM, in the finished product were performed.

The line included a dicer that reduced the particle size of the callaloo leaves and stalk in a manner similar to what was usually done manually. The product was then conveyed to an initial washer to remove the dirt (if any) and initial insect load on the plants. The rotation speed of the washer was 9 rpm and the time for complete callaloo washing and exit from the washer was set to not exceed 5 min to reduce leaf damage. The final washing was subsequently done by spray washing with potable water while the callaloo was on the conveyor. The results of these trials are presented in Table 8.5.

Table 8.5 **Efficacy of mechanized washer on XM Levels in callaloo** (*Amaranthus* sp.)

Trial #	Sample #	Whole larvae number (size in mm)	Whole insect number (size in mm)	Insect parts number (size in mm)	Other number (size in mm)
1	Raw	2 (2 and 1)	2 (2 and 1)	Nil	Nil
	Raw	2 (2 and 4)	1 (1)	1 (1)	Slug (2)
	Washed	Nil	1 (2)	1 (1)	Nil
	Washed	2 (2 and 5)	Nil	1 (1)	Nil
2	Raw	1 (1)	1 (1)	Nil	Nil
	Raw	1 (1)	Nil	2 (1 and 0.5)	Nil
	Washed	Nil	Nil	1 (0.5)	Nil
	Washed	Nil	Nil	1 (1)	Nil

Assessment of the results (Table 8.5) showed that although the processing system was able to reduce the insect load somewhat (raw vs. washed), it was ineffective in reducing the final XM counts any more than the traditional method used at contracted processing plants. The mechanical washer allowed residual whole insects, larvae, and insect parts, which, while below the limits (see further), were above what would be typically recovered with the improved agronomic practices and human intervention at the contracted facilities. As a result of this, the TF decided that the traditional approach to production in combination with the improved agronomic and field husbandry practices was a more effective production process (Fig. 8.5), and an improved and more stringent QA program could consistently deliver the results required.

Developing acceptable operating limits

In March 1994 the technical team had received confirmation of the level of XM that would be accepted in canned spinach. The letter from the FDA stated that "*the United States allows "defects" in canned spinach as follows; Aphids ≤ 50 per 100 gm, Larvae ≤ 2 × 3 mm each per 100 gm and Leaf Miners ≤ 8 per 100 gm. The letter continued by stating that there were no stated defect levels for e.g. collard or turnip or mustard green. There is no stated defect levels for callaloo, which is not classified.*" Information was also received as to which pesticides were registered for use with spinach and what the MRLs were. It became evident at that time that the list of specific pesticides for which the FDA-allowed residues excluded most of those currently being used with callaloo, vindicating the approach taken to explore new pesticides rather than malathion and Basudin, which were being used at the time. Fortunately, as had been determined earlier, the pesticides being used in the trials that were being completed were allowed on leafy greens (including *Amaranthus* sp.) and spinach, which the FDA had already indicated it would accept as a proxy for callaloo.

From the perspective of the TF, the tolerance for insect/insect parts for leafy greens, which was higher than for spinach, was more advantageous to be used for callaloo. However, based on the satisfactory results for pesticide residues obtained from the recent trials, subject to the managing/supervision of the callaloo farms in accordance with the SOPs, the production system would also be able to meet the standards for canned spinach. This, however, had to be done with controls in place to ensure that the protocols established, particularly the newly inserted step of brine washing to reduce the insect burden on the plant, were strictly adhered to, in addition to the strict monitoring of farm PHI management by the receiving contract packing firm. Once all of these were adhered to, the modified traditional production process could consistently deliver compliant product. While an assessment had been done on mechanization, based on the cost/benefit analysis undertaken, the greater production efficiencies delivered by a more mechanized production process did not offset the higher capital and recurrent operational costs. Neither did the mechanized process deliver a product with a lower insect burden. This made it more practical to continue to use contract packing arrangements for the production of callaloo to target delivery of a product that would comply with the DAL of canned spinach, as well as the applicable pesticide residue MRLs for Lannate, Depel, and Thuricide.

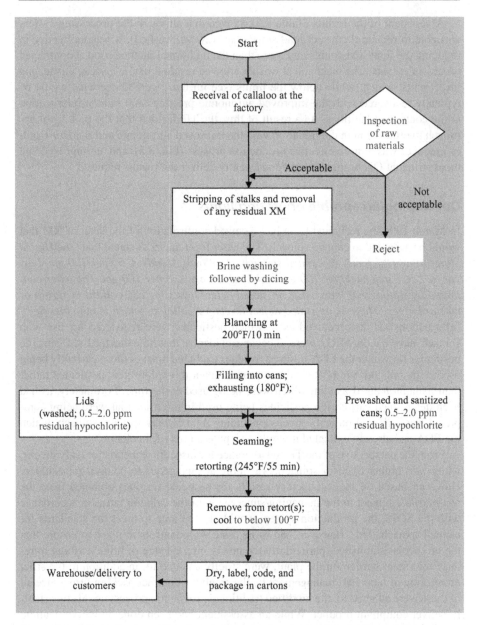

Figure 8.5 Adjusted process flow for production of canned callaloo (*Amaranthus* sp.).

The quality assurance program

The QA program for the production of canned callaloo was based on the process outlined in Figs. 8.2 and 8.3 and monitored the parameters indicated in Table 8.1. With the changes made to the agricultural protocols, the program extended back to the farms and consisted of the following:

- A GAP program based on the SOPs developed for the farming operations.
- Periodic audits of contract farming operations.
- Review of farm data, particularly pesticide application, and harvesting data (PHI monitoring).
- Periodic (monthly) audits of contract packaging operations, inclusive of in-line sampling
- Batch-by-batch pesticide residue testing.

In addition to these changes, after several trials and the assessment of the outcome of preprocessing handling and preparation, the research team was able to confirm that besides good postharvest handling and transportation practices, the most effective way to reduce the insect burden on the in-coming callaloo was to have the crop undergo a *brine wash* process. This was shown to loosen any residual insects that may have remained on the plant and allow for their separation from the plant before it was further prepared and, eventually, canned. This was therefore introduced as a mandatory step in the processing of callaloo for export (Fig. 8.5). The requirement for monitoring incoming product and checking the records was also improved to ensure adherence to the PHI.

In addition to increasing the stringency of practices on the farm and raw material receival and handling, the QA program was also improved to be more specific as to how selected aspects of the processing were handled and documented. For the washing and sanitizing of cans and lids, the minimum residual hypochlorite was specified and for cooling, the temperature to which cooling should be done (i.e., the water temperature) was also specified. Important changes to the process were denoted in color as shown in Fig. 8.5. In addition, raw material was sampled and checked for microbiological quality on a bimonthly basis (to monitor the use of spoiled callaloo for canning), the effectiveness of blanching checked on one in four batches by peroxidase activity, and the XM standard changed in accordance with the FDA standards for spinach (Table 8.6). The other change was the requirement that processors should monitor the effectiveness of the brine washing to ensure optimal removal of insects or insect parts. This program proved to be effective overall and was able to ensure consistent conformance with the required standards.

Petitioning the FDA to lift the import alert

Canned callaloo from Jamaica had not been offered for sale in the United States since the prohibition for over a year and so it was felt that the TF should ensure that everything was in place before it attempted to have trade recommenced. Having completed all that was required to comply with the XM (filth) DAL for spinach, as well as the applicable pesticide residue MRLs for both spinach and leafy greens, a dossier was

Table 8.6 **QA characteristics for canning of callaloo in 1994**

QA characteristic	Value/limit(s)	Comments
Raw materials		
XM	Absent	Visual assessment at plant
Microbiological quality (CFU/g)		
TAPC	<500,000	—
Coliform count	<1,000	—
Yeast and mold count	<10	—
Color	Dark green	—
Finished product		
Microbiological quality		
Commercial sterility at 35°C	Sterile	Sometimes done at 55°C
Sodium chloride salt (%)	1.0–1.5%	—
Vacuum (mm Hg)	127	—
Headspace (mm)	5–8	—
Texture	Soft, but not too soft	—
Drained weight (oz.)	15	—
Peroxidase activity	Negative	—
Can seam analysis	Various measurements	Canadian standards
Leaf to stalk ratio (70/30)	—	—
Stems < 7 mm	—	—
Leaves 60–70% when drained	—	—
Stems 30–40% drained	—	—
XM	Absent	Operating standard; FDA standard for canned Spinach: regulatory standard
In process		
Visual monitoring of brine washer efficacy	Insect burden significantly reduced; check residue in screen	—
Time–temperature monitoring	245°F/55 min	—

TAPC, Total aerobic plate count.

put together for presentation to the FDA as part of a petition to have the import alert removed in July 1994. It comprised of the details of the studies undertaken inclusive of the pesticides examined, the data obtained, and the experimental designs and statistical analyses of the results. These were shared with the FDA and its USDA counterparts with whom discussions had been ongoing. This provided them with data with which to assess the likelihood of ongoing compliance of the Grace-branded canned callaloo (and others made under the same regime) coming into the United States from Jamaica with the agreed tolerances for filth (XM) and pesticide residues. The petition requested that the DAL to be applied should be that for leafy greens, instead of spinach as leafy amaranth was already under that heading and callaloo was a leafy amaranth.

After consideration, the request was accepted and the standards for leafy greens applied. The import alert was then lifted on the successful delivery of five clean, compliant shipments to US ports of entry, accompanied by evidence of their compliance with the XM and MRL standards.

Conclusions

The Jamaican firm, GK and other Jamaican exporters were forced in 1993 to deal with the prohibition of exports of a canned traditional vegetable, callaloo (*Amaranthus* sp.) to the US market for breach of its foreign matter standards. In the process of trying to get the import alert that had been imposed removed, a TF had to address the ambiguity of the standard applied, ensure the improvement in agronomic practices for the crop, upgrade the process, and arrive at an agreement with the regulators in the United States as to the specific standards to be applied. The process to regain access was successfully completed in 13 months. It required monitoring the vegetable at all stages from farm to market to ensure the delivery of a product that complied with the DALs for insects and insect parts and the maximum residue limit (MRL) for pesticides approved for use by the EPA on the product.

Case study: contamination of callaloo (*Amaranthus sp.*) with pesticides residues

Background to the case

The industry in Jamaica had gone through the challenge of dealing with the import alert on callaloo (*Amaranthus* sp.) in 1993–94 and, having recovered, the major processors and exporters maintained the agreed practices up until the late 1990s into 2000. Over time, the agreed practices drifted and the memories of the challenges of 1993–94 faded. In addition, the lessons learned at that time about the importance of maintaining proper field husbandry practices and ensuring that only approved pesticides were being used were forgotten and the pesticides being used were changed without revalidating their suitability for use. This resulted in high levels of pesticides being detected for some samples of callaloo assessed by the FDA, which were being exported to the United States from Jamaica in 2000. This raised concern about the industry's willingness and ability to stick to agreed protocols (submitted to the FDA in 1994), use only approved pesticides, and abide by the limits on the crops being shipped to the United States.

The FDA reimposed an import alert on callaloo from specific firms entering the United States from Jamaica, a part of which was a strict testing regime for all shipment subsequent to the initial violative ones. The FDA issued an advisory stating that it would not accept any shipment of callaloo without accompanying certification that the vegetable complied with the acceptable types and level of pesticides approved for use on *Amaranthus* sp. in the United States. This began another round of interface with the US authorities on callaloo, but the issue was not XM as in the first case study (discussed earlier), but that the pesticide residues themselves exceeded the MRL and the pesticides being used at the time on the vegetable were not approved for use on *Amaranthus* sp.

The import alert

The import alert was imposed on exports of canned callaloo from the firm to the United States. This was reinforced by a letter sent from the FDA to the manufacturer. It is reproduced in its entirety (except for some names that are deleted) so that how these matters have to be dealt can be appreciated.

FDA Letter to Canned Callaloo Manufacturer regarding Pesticides

Dear Madam/Sir:

This is to inform you that the United States Food and Drug Administration (FDA) has examined a sample of canned callaloo in brine shipped by your firm to XXX Company,[n] Miami, Florida, on January 13, 2000. Analysis of the sample revealed residues of the pesticide chemicals, Diazinon, Chlorpyrifos, and Profenophos, which are illegal under the Federal Food, Drug, and Cosmetic Act, Section 408 (a).

Effective immediately, DCA is requiring that any shipment of canned callaloo in brine made by your firm to the United States be analyzed for the pesticides, Diazrinon, Chlorpyrifos, and Profenophos, as a prerequisite for entry. The analysis must be performed by a private laboratory either in the United States, or in your country, unless the government authorities in your country are willing to certify that the shipments now meet our requirements. Any shipment of canned callaloo in brine that is shown by certificate of analysis to contain illegal residues of Diazrinon, Chlorpyrifos, and Profenophos will be refused entry into the United States. Any shipment for which a certificate of analysis is not furnished will be automatically detained by FDA until an appropriate certificate of analysis is provided.

To enable us to evaluate your results of analysis and to approve release of further shipments, we recommend that the laboratory performing the analysis follow FDA guidelines for private laboratories, which are available on request. All analyses must be conducted by validated analytical methodology. Methods commonly used by the FDA may be found in the Pesticide Analytical Manual (PAM), volumes 1 and 11, and the Association for Analytical Chemists (AOAC) Book of Methods. The method of analysis used by FDA to determine Diazinon, Chlorpyrifos, and Profenophos in canned callaloo in brine was PAM, volume #1, Section 232.2. A copy of this is available upon request.

A full review of your analytical results may not be possible unless the certificate of analyses contains the information specified in the FDA guidelines for private laboratories, including among other things, a description of the sample collection, complete details on the analysis including the chromatograms, the address of the laboratory, and a statement of the qualifications of the analyst (s) who performed the analyses. FDA's guidelines also include forms for reporting the sample collection information and the details of the analyses. We suggest that these forms be used for the certification of shipments.

FDA will monitor the validity of the certificates of analyses by collecting periodic audit samples from randomly selected shipments of the product and testing them for the pesticide in question.

We will consider removing the requirement for certificates of analyses when, in our judgement, you have demonstrated that the residue problems no longer exist or have provided to FDA information that the canned callaloo in brine are not treated with the pesticides in question.

Data that would be useful to FDA in making a determination could include identification of the pesticide (s) used, dates and methods of their application and results of analyses from a representative sampling of the fields, reflecting that products from these fields are in compliance.

Please advise this office in writing within fifteen (15) days of the steps you will take with regard to this certification requirement and steps you will take to prevent similar pesticide problems in the future.

Please contact XXX, compliance officer, at (305) 526-XXXX,[o] ext. 929, if you have any questions or wish to discuss this matter further.

Cause of the problem

The problem that caused the imposition of the import alert, as is evident from the FDA communication previously, was the detection of diazinon, chlorpyrifos, and profenofos by the FDA on canned callaloo imported into the United States from Jamaica. Diazinon (0,0-diethyl-0-(2-isopropyl-6-methyl-4-pyrimidinyl) phosphorothioate), chlorpyrifos (0,0-diethyl-0-3,5,6-trichloropyridin-2-yl phosphorothioate, $C_9H_{11}Cl_3NO_3PS$) and profenofos (4-bromo-2-chloro-1-[ethoxy (propylsulfanyl)phosphoryl] oxybenzene, $C_{11}H_{15}BrClO_3PS$) are all organophosphate pesticides (Fig. 8.6), which were not approved for use in the United States on crops, such as spinach, leafy greens or *Amaranthus* sp. (callaloo). The EPA and the FDA had been reviewing concerns about the safety of organophosphate and carbamate pesticides for some time, with more than 300,000 fatalities globally being reportedly attributable to both each year (Bird et al., 2014). Based on these reviews, diazinon was banned in 2000 and the EPA banned the use of chlorpyrifos in 2001 (Bird et al., 2014), with other organophosphates being banned by 2003. The presence of these unapproved pesticides in the canned callaloo, therefore, drove the FDA to take action to prohibit further imports

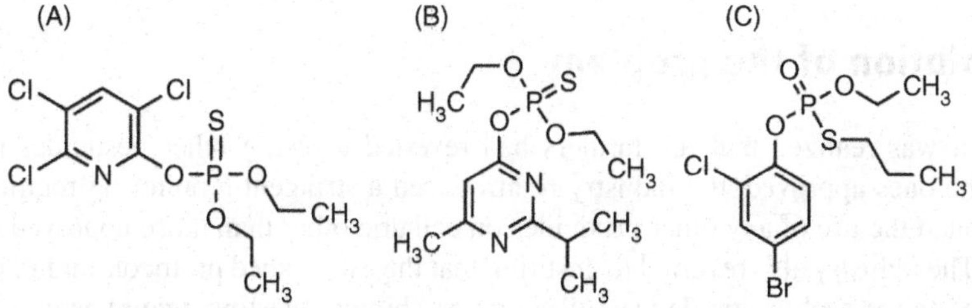

Figure 8.6 Structure of the organophosphates prohibited by the FDA on callaloo.
(A) Chlorpyrifos, (B) diazinon, and (C) profenofos.
Source: www.sigmaaldrich.com

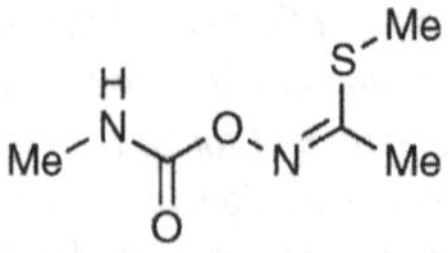

Figure 8.7 Structure of Lannate (a carbamate).
Source: www.trc-canada.com

until it could get sufficient evidence that the pesticides were not being used and were not present in the canned product at a detectable level.

The imposition of another import alert on canned callaloo created real consternation for manufacturers of the product in Jamaica who remembered having to forgo exports for over a year in the early 1990s. This was largely because the processor whose product was subject to the import alert, also supplied the same product to other suppliers, which was later found[p] to breach US regulations. Further, as discussed under the first case study, Lannate, Depel, and Thuricide were the only ones that had been approved for use on callaloo. It was critical, therefore, to find the root cause of the problem and address it quickly.

Lannate, although a carbamate (see structure in Fig. 8.7), was approved for use on *Amaranthus* sp. by the FDA. Along with Depel and Thuricide, both *Bacillus thuringiensis*–based pesticides, FDA approved the use of Lannate on the imported product, having gone through the process of formally approving these as discussed in the earlier case study. However, it appeared that after over 4 years of strict adherence to the agreed protocols, the farmers had drifted away from the use of the prescribed pesticides and started used chlorpyriphos (known by the trade name Dursban locally), along with the other organophosphate pesticides for treating callaloo. Furthermore, the processors and exporters had also become less stringent in their testing regimes and even more so in the monitoring of the farms, and those who routinely tested in lieu of farm visits/audits were testing for Lannate and not for the organophosphates being used. This meant that their analyses would not have picked up the presence of these pesticides in the crop at levels in excess of the MRL (<0.05 ppm, the detection limit of the equipment at the time), which was essentially zero (complete absence of these pesticides).

Resolution of the problem

Once it was realized that the farmers had reverted to using other pesticides rather than the ones approved, the industry reintroduced a stringent monitoring regime and prohibited the use of any other pesticides on callaloo other than those approved at the time. The industry also reverted to ensuring that the established protocols and PHI was adhered to for each pesticide used. Routine preshipment testing again became a part of the export process, with two established local laboratories[q] performing analyses on the batches of callaloo being produced prior to their export to the United States. The results of the findings were used both to guide the farmers as to the effectiveness of their husbandry practices and also to eliminate those who would not adhere to the guidelines.

This increased monitoring of the farms and analyses of the product prior to export resulted in drastically reduced levels of pesticides on the product to the established MRLs range and facilitated the resumption of export of callaloo into the United States under the terms stipulated by the FDA. After a period in which between five and eight consecutive shipments from each exporter were shown to be consistently clean, unhindered exports resumed.

Conclusions and lessons learned

Both of these case studies demonstrated the importance of having control over the entire supply chain when it comes to ensuring the compliance and safety of food products. This is being emphasized and enshrined in the current US law through both the requirements of the FSMA rules commonly called the Produce Rule, the Preventive Controls for Human Foods Rule, and the Foreign Suppliers Verification Program (FSVP) Rule [Gordon, 2015a; Food and Drug Administration, 2015a,b,c; Food Safety Preventive Controls Alliance (FSCPA), 2016]. The first case study demonstrated the importance of collaboration along the value chain in solving technical problems that arise, as well as effective communication and, where possible, collaboration with the regulators of the importing country. The second case study showed that although industry value chains often get affected by market access interruptions for various reasons, they often forget the problem after a few years. The important lesson is that constant vigilance and the implementation of a system of monitoring and enforcement of compliance to standards needs to be a part of any sustainable solution to a market access challenge that is solvable by the application of a food science–based technical solution.

The challenges with pesticide residues and XM and their interrelationships, which were explored in this chapter, are relevant for all horticultural exports that rely on the use of prophylactic methods, which help in the control of pests. While many years have passed since these challenges with the export of canned vegetables to the United States arose, the situation regarding rejection of products at port of entry has not changed significantly (Gordon, 2015a). The regulations have been updated but the essential nature of the problem remains, making the cases studies presented herein very relevant to the approaches that can be applied.

Endnotes

[a]*Amaranthus viridis* is called "callaloo" in Jamaica and, by extension, in the diaspora populations in North America and the United Kingdom. In the rest of the Caribbean, the leaves of the dasheen (*Colocasia esculenta*) are called callaloo. Both are used for similar culinary purposes. Dasheen is also widely known as *taro* in Latin America and the United States.
[b]Collard greens are part of a group of vegetables collectively known as "greens" in the United States. Although a part of the species *B. oleracea*, which also includes cabbage (Capitata group)

and broccoli (Botrytis group), collard greens are distinguished by the membership in the Acephala group, which includes spring greens and kale.

[c]Defect action levels (DALs) are limits set by the Food and Drug Administration (FDA) to define the extent of contamination acceptable in food. The action level represents the limit at or above which FDA will take legal action against a product to remove it from the market for being adulterated ("unfit for food").

[d]Violative means noncompliance with the FD&C Act or associated regulations enforced by the FDA.

[e]Detention without physical examination (DWPE) means that the regulated food or drug item is being "refused admission by the FDA based on information, *other than the results of examination of samples*, that causes an article to appear to violate the FD&C Act."

[f]Postharvest interval (PHI) is the period postapplication after which the residue levels on specific crops would be within the allowable pesticide maximum residue level (MRL) for that pesticide/active ingredient.

[g]The technical support team included the author and other senior team members of the Grace Technology Centre (GTC), Technological Solutions Limited's (TSL) precursor entity and the R&D, quality assurance (QA), and technical extension support center of GraceKennedy (GK) at the time.

[h]The second case study examines what transpired when breaches in PHI protocols and integrated pest management (IPM) caused this to become a reality.

[i]This requirement was a part of the upgraded QA program, discussed further, to ensure some level of supplier control on receipt of raw materials. It is interesting to note that supplier controls, such as these (being enacted in 1993–94), are now a requirement under the FSMA (21 CFR 117).

[j]It was important to have representation of key stakeholders at this discussion to ensure that nothing critical was overlooked. Those involved were Mr. William Fielding [Ministry of Agriculture (MOA)], Ms. Jean Shand (GK agronomist), Dr. Ian Thompson (GTC), Mr. Ewan Findlay (GK engineer), Dr. Brian Davidson (GK export division), and Mrs. Arline Blackwood (GTC).

[k]Code of Federal Regulations Title 40 (CFR 40) deals with the Environmental Protection Agency (EPA) and the applicable pesticides for different crops, along with detailed applicable regulations.

[l]Analyses were done by the Pesticide Residue Laboratory at the University of the West Indies, the only laboratory equipped to do confirmations at that time (1993). These tests were used to confirm GTC's findings using test kits.

[m]Now called Campden BRI. Campden is one of the most respected food science and technology research centers in Europe and one with which GK had a long-standing relationship.

[n]Company name withheld.

[o]Name and actual telephone number withheld. All other aspects of the letter were exactly as presented.

[p]TSL was called in on the receipt of the import alert notice by the processor and associated large export trading houses to address the problem.

[q]TSL and the University of the West Indies Pesticide Residue Laboratory in the Department of Chemistry.

Case study: addressing the problem of *Alicyclobacillus* in tropical beverages

A. Gordon
Technological Solutions Limited, Kingston, Jamaica

Chapter Outline

Introduction

In an increasingly globalized world where the nature of the foods being consumed and sources of supply are becoming more diversified, fruit juices, juice drinks, and fruit-based beverages occupy an important role. Not only are these a major source of hydration, they, along with vegetables, are often one of the major sources of an important mix of vitamins and minerals such as vitamins A, B, C, D, E, and K, calcium, potassium, sodium, selenium, and zinc, among others, required for optimal functioning of the body. Tropical fruits such as acerola (*Malpighia emarginata*), also called West Indian cherries, guava (*Psidium guajava*) and lychee (*Litchi chinensis*) from developing countries are very rich in vitamin C (1.6 mg/g, 2.28 mg/g, and 1.83 mg/g, respectively), as are more traditional fruits in developed country diets such as oranges

Food Safety and Quality Systems in Developing Countries. http://dx.doi.org/10.1016/B978-0-12-801226-0.00009-8

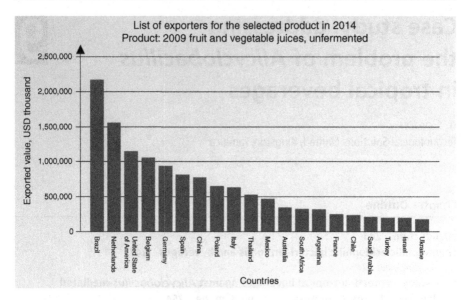

Figure 9.1 Global exports of fruit and vegetable juices, 2014.
Source: ITC Trade Map.

(*Citrus × sinensis*) 1.38 mg/g and black currants (*Ribes nigrum*), 1.8 mg/g. Others, such as carrot (*Daucus carota*) and beet root (*Beta vulgaris*) juices are excellent sources of vitamin A (334% of the recommended daily value) and potassium (3.20 and 3.25 mg/g), respectively. These beverages are therefore an important component of global diets, representing a US $12.52 billion industry, with Mexico, Argentina, Thailand, Chile, and South Africa being major players (Fig. 9.1). In addition, the changing tastes of consumers in developed country markets (Gordon, 2015a) is driving innovation that has fuelled an explosion in the range and variety of tropical fruit and vegetable juices, juice blends and juice drinks available in the distribution channels throughout the world (Fig. 9.2). The combination of these factors, as well as the

Figure 9.2 Examples of fruit and vegetable juices and juice drinks on selected developing country markets.
Sources: Barbados Dairy Industries Limited, http://forums.redflagdeals.com/where-buy-grace-tropical-rhythms-juice-jarritos-pop-1739801/ and Lasco Manufacturing Limited.

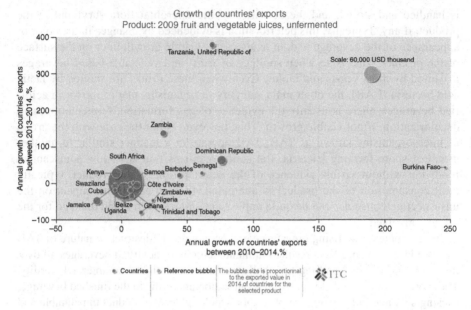

Figure 9.3 Rate of growth in fruit and vegetable juice exports from developing countries 2010–14.
Source: ITC Trademap (from the COMSTAT database).

trend toward more healthy eating and healthy lifestyles has meant that the contribution of fruit and vegetable juices, juice blends, and juice drinks to global diets and lifestyles continues to have a major impact on the food industry in all markets.

The global demand and trade in fruit and vegetable juices is increasing, with developing countries such as Barbados, Senegal, and the Dominican Republic growing at rates of between 45% and 80% over the period 2010–14 and with Zambia (130%) and Tanzania (380%) growing their exports very rapidly in 2014 (Fig. 9.3). This rapidly increasing demand presents continuing opportunities for growth of the beverage industry in producer countries as consumers in developed country markets want to be able to get their nutrition from an expanding array of novel and enjoyable tropical fruit beverages. Consequently, any threat to the acceptability and exportability of these beverages could not only impact supply in the marketplace, but can also affect the economy of the significant exporter countries and the individual firms involved. This is the context in which the impact of the role of thermophilic acidophilic bacteria (TAB) in the beverage industry in developing countries should be viewed.

Thermophilic acidophilic bacteria in foods and beverages

All beverages deteriorate with age, with the exception of wine and certain distilled liquor. The rate of deterioration depends on the storage temperature, the composition of the beverage, type of container used, light absorption, the way the product is handled and stored, and the activities of the microbial flora surviving in the product, if any. Typically, this deterioration is evidenced by changes in the color or appearance of the beverage and, in many cases, visible growth in or on the surface of the product. This is so when spoilage of fruit- and vegetable-based beverages is caused by most yeasts and molds. Even when these drinks are spoiled by lactic acid bacteria (LAB), the other main category of organisms that can grow in acidi- fied beverages, there is usually the evidence of gas formation, fermentation, and decolorization, if not visible growth. This, however, is not the case with the group of microorganisms known as TAB. TAB operate in a manner similar to another group of spore-forming bacteria, flat sour organisms that spoil low acid canned foods but without visible evidence of the deterioration of the product, typically until consumption of the product is attempted. In both cases, the impact on the unsuspecting consumer can be quite unpleasant and result in brand damage for the product involved.

In this context, the rising prevalence and economically destructive nature of TAB over the last 2 decades have resulted in manufacturers of acidified beverages all over the world faced with handling the challenge to their products' commercial sterility. These organisms are able to survive pasteurization and grow in the finished beverage, causing spoilage, off-flavors and off-odors which render the product unpalatable and undesirable to consumers, without any prior indication to warn the consumer that the quality of the beverage has been compromised. Pasteurization is the process by which sufficient heat is applied for a predetermined period of time such that microorganisms of public health significance or those which traditionally grow in beverage products are destroyed and the beverage made commercially sterile.[a] The pasteurization cook that is applied to beverages is intended to free the products from any microorganism that may cause their deterioration or affect the health of the consumer. Unfortunately, while the thermal processes commonly used are able to destroy pathogens such as *Escherichia coli* O157:H7 and *Salmonella*, they are not effective against the spore- forming spoilage TAB (Splittstoesser et al., 1994) of which the genus known as *Alicy- clobacillus* is the most typical. Thermal processes that able to destroy or incapacitate *Alicyclobacillus* spores are not commercially viable as they are typically so severe that product quality is irreparably compromised (Walls and Chuyate, 1998). As a result of this conundrum, other ways have to be found to deal with the TAB organisms which, if left unchecked, will render significant production within the juice and juice drink segment of the beverage sector unsalable. It is therefore important the participants in the fruit juice, juice drink, and fruit-based beverage value chain understand the nature of the threat and effective ways to deal with it.

This chapter provides a detailed case study of *Alicyclobacillus* contamination of a tropical beverage in a production plant setting and the approaches applied in control- ling and then eliminating the organism as a contaminant to the process. Based on the detailed reviews of the topic done by Ciuffreda et al. (2015), Tianli et al. (2014), and Chang and Kang (2004), the case study presented here is one of the first reported cases of a successful approach to addressing *Alicyclobacillus* contamination in a commer- cial production setting using industrially relevant approaches in a developing country context.

Alicyclobacillus

Nature of the organism

Alicyclobacillus is a genus of rod-shaped, Gram-positive bacteria that is usually recovered from soil, orchards, acidic drinks, and equipment from fruit juice manufacturers. They are spore-forming, thermophilic microorganisms that are obligate aerobes (Ciuffreda et al., 2015). The genus consists of a group of thermoacidophilic, strictly aerobic, heterotrophic, and spore-forming bacteria (Walls and Chuyate, 1998; Wisotzkey et al., 1992) and belongs to the family of Alicyclobacillaceae (da Costa and Rainey, 2010). The spores of *Alicyclobacillus* are able to survive typical pasteurization of juices (at 92°C for 10 s). The vegetative cells grow well between 42 and 60°C and in substrates or products with a pH 3.5–4.5. The genus consists of several species including *Alicyclobacillus acidoterrestris*, *Alicyclobacillus acidocaldarius*, *Alicyclobacillus pomorum*, and *Alicyclobacillus acidophilus*, all with varying characteristics, the details of which are explored in subsequent sections. *A. acidoterrestris* is the main organism implicated in the spoilage of beverages, although *A. acidocaldarius*, *A. pomorum*, and *A. acidophilus* have been isolated from spoiled products and can also produce compounds that taint beverage products. *Alicyclobacillus*, while not considered a food safety hazard, is therefore a quality control target organism for manufacturers of acidic beverages because of the deleterious effect its presence can have on the aroma, flavor, and hence the ultimate quality of fruit drinks and juices.

Extent of the problem and sources of contamination

The first documented case of spoilage by bacteria of this genus was recorded in Germany in 1982 in pasteurized fruit juice (Tianli et al., 2014). Since that time, because of the products targeted, the organism is now generally regarded as one of the major causes of the spoilage of acidic fruit drinks and related beverages. A survey done in the USA in 1998 by National Food Processors Association (NFPA)[b] reported that spoilage caused by acidophilic spore-formers was experienced by 35% of fruit juice manufacturers (Chang and Kang, 2004; Walls and Chuyate, 1998). Other studies have indicated that processors who have reported spoilage incidents in fruit juices range from 45% in Europe to 24% upward for Argentinian fruit beverages (Tianli et al., 2014). Spoilage is typically suspected to have been caused by *A. acidoterrestris* in most cases. Other beverages that have experienced spoilage by these organisms included carbonated fruit juice drinks, white grape juice, canned diced tomatoes, apple, pear, peach mango, and orange juice (Oteiza et al., 2011, 2014; Tianli et al., 2014). In single-strength juice, these microorganisms find a favorable environment for germination, growth, and spoilage (Chang and Kang, 2004). In a recent review of the status of *Alicyclobacillus* contamination and spoilage of fruit juices, Ciuffreda et al. (2015) indicated that there has been a significant increase in the cases of spoilage of pasteurized fruit juice products by *Alicyclobacillus* spp. in the last few years (Bevilacqua et al., 2008a; Pettipher et al., 1997; Silva and Gibbs, 2004; Walker and Phillips, 2008; Yamazaki et al., 1996).

Alicyclobacillus spp. may be present on fruit surfaces contaminated by soil during production and harvesting (Eiroa et al., 1999). It is thought that contaminated fresh fruit introduced during processing without proper cleaning could be a major source of contamination and subsequent spoilage for juices made from fresh fruits (Pontius et al., 1998; Splittstoesser et al., 1994). Contamination may also be caused by the use of unwashed or poorly washed fruits during processing. The spores may also have been introduced into the facility by the soil employees take in on their shoes or person. The water used during processing may be another source of contamination if it contains the spores or vegetative cells of the alicyclobacilli (Tianli et al., 2014). Knight and Gordon (2004), in a study of tropical fruit juices and juice drink blends contaminated with *Alicyclobacillus* while suspecting the initial source of the contamination to be the concentrates used in the juice drink blends were unable to find specifically identify the source of the spores. They did, however, find the organism in tomato paste which had been processed into ketchup using some of the same equipment. This was in keeping with the findings of Walls and Chuyate (1998) who found the organism in tomato products. Other studies of juices and acidified beverages have found that fruit juice concentrates, fruit pulps, and flavorings (Oteiza et al., 2014) were implicated as the source of contamination. Alicyclobacilli have been found in multiple areas and surfaces within the food processing environment of beverage plants (Zhang et al., 2013). What is clear is that failure to identify the source of the contamination may result in the establishment of a persistent infection throughout the operations (Knight and Gordon, 2004) that will continually run the risk that the production and sale of high quality product that will have an acceptable shelf life and quality will be compromised (Ciuffreda et al., 2015; Tianli et al., 2014). This has been the result in products ranging from juices produced in both tropical and temperate countries including apple, pear, peach, and grape (Chang and Kang 2004) to isotonic water and lemonade, carbonated fruit juice drinks and iced tea with fruit juice which were found to contain *A. acidoterrestris* (Pettipher et al., 1997; Yamazaki et al., 1996).

Alicyclobacillus does not produce gas when it metabolizes sugars, but instead produces acids which make it difficult to detect. In some cases, spoiled juices experience little or no pH changes and will appear normal with occasional turbidity development at the bottom of the container (Ciuffreda et al., 2015). The impact of the presence and action of *Alicyclobacillus* on beverages is most typically characterized by the off-odor associated with guaiacol (2-methoxyphenol) and other halophenols. (Ciuffreda et al., 2015). These are the by-products of the metabolism of the organism in fruit juices and juice drinks. Some ingredients in these products, including vanillic acid and tyrosine, provide substrates for guaiacol production. The nature of the outcome of the metabolic activity of the organism in the product is dependent on the concentration of these and related substrates, the availability and concentration of oxygen in the beverage, storage temperature, the initial viable count, and the application of heat as part of the thermal processing of the product during pasteurization which potentiates the outgrowth of the alicyclobacilli spores (Chang and Kang, 2004; Pettipher et al., 1997). While all *Alicyclobacillus* species such as

A. acidoterrestris and *A. acidocaldarius* do produce acid as a by-product of their metabolism, they do not significantly affect the pH of the product and therefore may not be effective antagonists for the growth of other pathogens like *Clostridium botulinum* as LAB have often been shown to be (Okereke and Montville, 1991a,b; Shivsharan and Bhitre, 2013).

Phylogenetic classification, differentiation, and enumeration of Alicyclobacillus

Differentiation from other bacilli

Initially, alicyclobacilli were placed in the genus *Bacillus*, because they form endospores in a manner similar to other bacilli. However, phylogenetic analysis based on the comparisons of the 16S rRNA sequences showed that species of this genus belonged to a distinct line of descent within the Gram-positive lineage of bacteria with low guanine + cytosine (G + C) content. This low G + C lineage also included *Sulfobacillus* spp., a facultative autotroph (Wisotzkey et al., 1992; Tourova et al., 1994; Durand, 1996). In 1992, this group of organisms were allocated a new genus known as *Alicyclobacillus*, primarily due to the presence of ω-cyclohexyl or ω-cycloheptyl fatty acids as the major natural membrane lipid component (Oshima and Ariga, 1975; Hippchen et al., 1981). These ω-alicyclic fatty acids have been associated with the heat and acid resistance of *Alicyclobacillus* spp. (Chang and Kang, 2004), as they are responsible for the ability to survive typical pasteurization regimes applied during juice, juice drink and acidified beverage manufacturing. The alicyclobacilli also have some distinct biochemical and phenotypic characteristics that can facilitate differentiation from other bacilli in a commercial setting. These are discussed later in this chapter.

Differentiation within the genus

Although some studies suggest tolerance by some strains for low (0.1%) levels of oxygen (13), *Alicyclobacillus* spp. are typically strict aerobes. All members of the genus produce acid, without gas, from the metabolism of sugars, and all also produce guaiacol, although the quantity varies significantly between species (Chang and Kang, 2004; Goto et al., 2007; Bevilacqua et al., 2008b). The genus can be characterized by phylogenetic analysis and most display similarity in phenotypic expression. Differences in the latter, however, can be useful in differentiation of the various species, particularly characteristics such as acid production, the use of various carbon sources, starch, and gelatin hydrolysis, catalase and oxidase reaction, resistance to 5% NaCl and nitrate reduction which are species- and strain-dependent (Shemesh et al., 2014). Originally consisting of three species, *A. acidocaldarius*, *A. acidoterrestris*, and *Alicyclobacillus cycloheptanicus*, the genus now encompasses 22 species isolated from various sources around the world. Twelve of these including *A. acidocaldarius*, *A. acidoterrestris*, and *Alicyclobacillus acidiphilus* contain ω-cyclohexane fatty acids (Wisotzkey et al., 1992; Matsubara et al., 2002), *A. cycloheptanicus*, and four others contain ω-cycloheptane fatty acids (Deinhard et al., 1987;

Goto et al., 2007), while recent studies have indicated that *A. pomorum* and four other atypical alicyclobacilli lack both fatty acids (Glaeser et al., 2013; Goto et al., 2003). The atypical members of the group were shown by genotypic analysis to be phylogenetically related to members of the genus *Alicyclobacillus* (Guo et al., 2009; Goto et al., 2003, 2007; Jiang et al., 2008). They did, however, differ in phenotypic characteristics from classical alicyclobacilli in their growth temperature, use of carbon sources and ability to utilize sources such as ferrous iron, tetrathionate, and elemental sulfur for growth (Guo et al., 2009; Goto et al., 2003, 2007). A more detailed description of the genus can be had from a recent review by Ciuffreda et al. (2015) and Tianli et al. (2014).

Thermal susceptibility and traditional approaches to handling Alicyclobacillus

Juices and juice drinks are generally pasteurized to eliminate or reduce any bacterial contaminants present in the product and increase their shelf life. Under 21 CFR Part 120 (generally called the Juice HACCP Regulation), the United States Food and Drug Administration (FDA) requires that the thermal process applied to fruit juices or juice drinks should achieve a minimum of a 5-log (5D) pathogen reduction, with the major target organism being *Escherichia coli* (US Food and Drug Administration, 2001). Typical pasteurization processes commonly used to destroy pathogens such as *E. coli* O157:H7 and *Salmonella*, utilizing temperatures of between 85 and 95°C, including hot-fill-and-hold processes, easily achieve this. Other regulatory bodies around the world have similar expectations, including under legislation such as the Safe Food for Canadians Act (2012)[c] and similar European Union Directives and Regulations. Pasteurization is the process by which the use of high temperature, short time (HTST), ultra high temperature (UHT), and other equivalent processes, followed by rapid cooling, are used to achieve commercial sterility while preserving the organoleptic and nutritional properties of the beverages. Unfortunately, these pasteurization processes are not effective against thermotolerant spore-forming spoilage bacteria (Splittstoesser et al., 1994), do not eliminate the spores and may instead potentiate their outgrowth. As it is a spore-former, *Alicyclobacillus* has been shown to easily survive typical thermal processing treatments applied to juices, juice drinks, and fruit-based beverages (Chang and Kang 2004; Silva et al., 1999).

Thermal processes able to destroy *Alicyclobacillus* spores are not feasible as they are likely to be very deleterious to product quality (Palop et al., 2000; Walls and Chuyate, 1998). *A. acidoterrestris* has been reported to have a D_{75-95} values ranging from 1 to 85 min in various substrates, including commercial beverages (Silva et al., 1999; Splittstoesser et al., 1994, 1997; Tianli et al., 2014). z-Values ranging from 7.2 to 10.8 have been reported (Silva et al., 1999), while other thermal susceptibility studies reviewed by Tianli et al. (2014) have reported z-values of 5–22°C, depending on the species of *Alicyclobacillus* and the thermal processing conditions. This creates a real challenge for beverage manufacturers faced with the threat of *Alicyclobacillus* and the need to ensure that they insulate their production processes against possible quality problems should they ever be exposed to the possibility of the organism being present. The concern is heightened as even relatively low numbers of *Alicyclobacillus* are able to cause spoilage and produce objectionable flavors and odors in the beverages, significantly reducing consumer acceptability and possibly damaging the brand.

The ability of *A. acidoterrestris* and other alicyclobacilli spores to survive the thermal processes used during fruit juice and juice drink production and the inability to use more destructive thermal process because of loss of quality creates a conundrum for processors. It therefore requires the design of alternative techniques to reduce, eliminate, or capture contamination of the product with these organisms (Eiroa et al., 1999; Splittstoesser et al., 1994), thereby allowing effective mitigating steps to be taken. The use of nonconventional approaches to control alicyclobacilli have been explored, with several of these providing differing levels of effectiveness in eliminating or preventing the risk of quality loss due to the bacteria. Some approaches proved to be very effective in a laboratory setting but have yet to be successfully employed in a commercial environment, as has been found in detailed reviews of the literature on the various approaches studied and the available options (Chang and Kang, 2004; Ciuffreda et al., 2015; Tianli et al., 2014). Consequently, the major challenge to the industry has been the application of techniques and approaches developed in the laboratory to an industrial setting in a manner that ensures consistent effectiveness in mitigating the threat of economic losses caused by *Alicyclobacillus*. The case study reported here addresses this issue with solutions geared which were applied in an industrial setting with options available to many beverage manufacturers.

As with their global counterparts, beverage manufacturers in developing countries have found themselves faced with this threat and have to find ways to deal with it. The remainder of this chapter details a case study in which the sporadic contamination of a range of juice drinks with *Alicyclobacillus* was addressed in 2002. While this was before the advent of some of the more recent tools that have become available, the technology and approaches used are still being used today. Working with the best available science at the time, the cause of the sporadic spoilage of the nonclarified fruit juice was identified first as TABs, then as atypical *Alicyclobacillus*, and interventions developed to eliminate the organism from the finished products. These included a range of approaches which remain relevant, including ribotyping, physicochemical, and biochemical methods, which were then translated into simpler approaches that can be used to identify whether the problem being experienced was likely caused by TAB. This is important as many of the options for dealing with this group of organisms are either not available in many developing countries or too expensive to be affordable to the firms involved. The approach involved refining methods to identify and enumerate the organisms and their metabolic by-products in juices, identifying the source of TAB in the product and within the production system, and examining the effect of ingredients, processing conditions, and mitigating interventions on their control. A quality assurance protocol specifically for early detection, elimination, and prevention of outgrowth of TAB, if present, was also developed and successfully implemented.

Case study: protecting tropical beverages against *Alicyclobacillus*-mediated spoilage and the resultant possible brand damage

Nature of the problem

GraceKennedy & Company Limited (Grace) is a major conglomerate involved in the production, distribution, and export of Caribbean foods to global markets. With subsidiaries throughout the Caribbean, in North and Central America, the United Kingdom and Western Africa, Grace prides itself on delivering the taste of the Caribbean to the world. Located in Kingston, Jamaica, GraceKennedy, manufactures and exports a variety of food products, including a range of beverages to North America, Europe, and the Caribbean. Grace had developed a very successful and growing line of tropical beverages, Tropical Rhythms, in several flavors, including Tropical Rhythms Fruit Punch (TRFP), Pineapple Ginger and Mango Carrot (Fig. 9.4). In the early 2000s, the sales of Tropical Rhythms had begun to grow rapidly, stretching its capacity to keep up with demand, expanding locally, regionally in the Caribbean, and internationally into North America and the United Kingdom. Coinciding with the rapid growth of the product, Grace found that like other acidified beverage manufacturers around the world, it began experiencing sporadic spoilage and off-flavor issues for this line of acidified fruit juice drinks. They tried to identify the source of the problem but were unable to understand how juice drink blends with an acid pH of less than 4.5, and which had also been pasteurized, were having issues with spoilage and off-flavors. This was further compounded when traditional microbiological and quality assurance checks which were routinely undertaken by the company failed to pick up any microorganisms to which the problem could be attributed.

Figure 9.4 Variants of Grace Tropical Rhythms Juice Drinks.
Source: http://www.tastyjamaica.com.

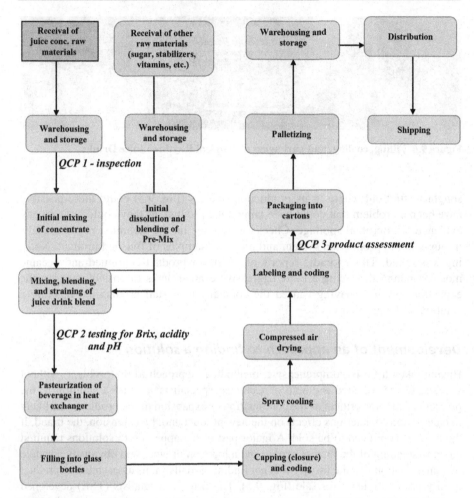

Figure 9.5 Outline of commercial processing of Grace Tropical Rhythms Juice Drinks prior to intervention. *QCP*, Quality control point.

Grace had developed a production process for the TRFP, the main product in the new lines of products, and other flavors (Fig. 9.5). It involved the preparation of the incoming raw materials into preblends which were then further blended as required to give the desired Brix, acidity, and consistency, prior to pasteurization at a minimum temperature of 92°C for a minimum of 30 s in a plate heat exchanger. This was then cooled in a cooling tunnel, prior to filling and packaging (Fig. 9.6). The entire system was PLC-controlled, ensuring consistency in both the pasteurization and cooling process. The advent of the problem first saw few, scattered complaints about the products not smelling as they should and tasting "medicinal." As most manufacturers would, when faced with this problem and without knowledge of the (at the time) relatively unknown group of organisms known as TABs, the firm

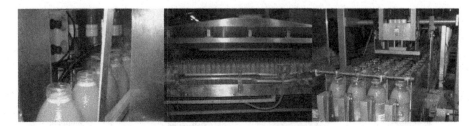

Figure 9.6 Filling, cooling, and packaging of Tropical Rhythms Juice Drinks.

sought to find out where in its production process (Fig. 9.5) could there possibly have been a problem that could have tainted the product, as they could not find any evidence of microbial spoilage. Checks were done on the incoming raw materials, throughout the production system and on many samples of finished product. Nothing was found. This sporadic reports of off-flavor product continued and became more pronounced, forcing a more aggressive evaluation as to possible causes, the expanding problem having caused the company to curtail production, pending a solution to the problem.

Development of an approach to finding a solution

Having taken a fairly comprehensive, methodical approach to the problem, without success, Grace solicited the help of technical specialists[d] to find a solution to the problem. Time was critical as the continuation or expansion of the problem was likely to have serious deleterious effects on the new product and, by extension, the brand, if the product continued to be sold. A major part of the approach to solutions required an understanding of the factors at play and a location of what was obviously a source of contamination that the facility was not finding initially on its assessment of product quality prior to release for sale (Fig. 9.5). The firm of specialists (TSL) developed an approach that required an even more detailed assessment of the raw materials and other inputs; the production process, including the thermal process and its efficacy; the microbiology of the product, in greater detail that was done before; and research into possible causes for the spoilage being experienced, particularly its sporadic nature. This was followed by the development of approaches to identify the critical points at which specific process parameters and product characteristics had to be controlled to prevent spoilage, as well as the institution of mitigating and preventative measures to eliminate the risk of spoilage or reduce it to an acceptable level for product which would be identified as being at risk prior to release for sale. This was critical to a two-step process which first saw the problem being controlled by early detection and mitigation, followed by the implementation of a permanent solution that has managed to eliminate the problem and prevent a recurrence over the last 14 years. The two-step approach was required to ensure that production could be restarted and the market kept supplied with Grace Tropical Rhythms juice drinks while a permanent fix was found.

The overall approach to facilitate a quick restart of production required the following:

- Identifying the specifics of the bacteria causing the problems: their sources, behaviors in the products, and raw materials of interest, the specific characteristics that could point to means of effective control, their distribution in the plant setting and other nuances that were peculiar to them.
- Developing and verifying appropriate methodology for isolating and enumerating these organisms, if present in raw materials and other potential sources of contamination (e.g., equipment).
- Identifying their metabolic products, if present, in raw materials and on equipment, thereby alternative and effective means of confirming their presence rapidly and precisely. This was to facilitate early detection and effective control, prevention or removal from the production chain.
- Identifying means of inhibition/prevention/elimination of the bacteria from the production process, specifically identifying any raw material that consistently introduced this contaminant into the process.
- Developing an appropriate quality control and assurance program that would reduce the risk of a recurrence of the problem to an acceptable minimum to facilitate a restart of production.

To address the problem permanently and instituting an effective prophylactic program, the following was required:

- Validating the effectiveness of the means being applied to destroy the organisms, should they be present in the in-process material or inhibit their growth, if it was not possible to prevent them from getting into the product.
- Developing and validating suitable techniques for predicting their growth in finished product and for detecting their presence in the production system whenever this arose, in a timely enough manner to take preventative action so that tainted product could never reach the market.
- Developing an approach to ensure that measures taken to eliminate the organisms from the production environment and the product was effective.
- Refining the quality assurance program to ensure that the risk of undetected contamination of any production lot was eliminated to facilitate confidence in the product in support of its an ongoing export expansion program.

To advance the process of successfully achieving the urgent objective of a quick restart of processing, initial detailed studies were undertaken that fully characterized the microbiological flora of each area of the production system as well as each of the inputs into the product and process. This required a detailed mapping of the process and reconfirmation of the efficacy of each step, as well as determining the possible causes of the spoilage, identifying the off-flavor characteristic of the spoiled product and determining why spoilage, when it did occur, was unpredictable and sporadic at the time (Knight and Gordon, 2004). Both a sensory and, more importantly, an instrumental method also had to be found to specifically identify and measure the off-flavor that was being experienced and relate this to specific process parameters or inputs into the process. As no one at the time knew what the problem was, the process therefore had to be dissected from start to finish to ensure that what appeared to be a fairly routine acidified fruit juice drink production process was fully understood as to how it could be resulting in this highly problematic sporadic failure of product quality.

The original production process

The production process for Grace Tropical Rhythms Juice Drinks as it was at the time is outlined in Fig. 9.5. Raw materials were received into storage, prior to an examination in the laboratory at the Grace subsidiary where the beverage was produced. The inspection and assessment assured that the inputs were free from extraneous matter, was within specification in terms of their microbial loads and met all other internally established specifications. When required for production, these inputs were blended in separate kettles with the dry ingredients undergoing dissolution and sieving prior to being mixed with the fruit juice and juice drink concentrates. The blended mixture was then pasteurized through a plate heat exchanger and filled on a rotary, multihead filler into 8 oz. glass bottles (Fig. 9.6). Filled bottles were then capped and cooled in a cooling tunnel under a constant spray of chilled water graduated in temperature from start to finish to eliminate heat shock for the hot beverage but ensure that the product was fully chilled by the time it exited the cooler. The fully cooled juice drinks were then dried, any low vacuum bottles removed with a dud detector[e] and the product labeled, warehoused and stored, prior to distribution and sale. The major point of control in the original process is identified in gray (Fig. 9.5).

Detection of spoilage organisms in fruit purees, concentrates, and tropical rhythms fruit juice drinks

Tropical Rhythms Fruit Punch (TRFP), that had been spoilt, fruit purees, concentrates, and other flavors of Tropical Rhythms were prepared for assay for the presence of spoilage organisms, and any bacteria found isolated and enumerated by diluting the samples with sterile distilled water to bring the Brix to within a range of 11–14. For samples with Brix measuring less than 11 no dilution was done. The samples (500 g) were heat shocked for 10 min at 80°C, cooled in an ice water bath, and then incubated at 43°C for 10 days. Following this incubation period, 8 mL of the samples were plated directly to Nutrient Agar and acidified K Agar, the latter being a differential medium for TAB. The plates were then incubated for 48 h at 35°C and 96 h (4 days) at 43°C, respectively, aerobically and anaerobically, and observed for growth. All samples plated onto K Agar were also evaluated olfactorily (by smelling) for the odor of guaiacol, a typical metabolite of *Alicyclobacillus*.

Detection of TAB in sugar syrup (60 Brix), cooling water, and on equipment swabs

For sugar syrup (60 Brix), 10 mL was heat shocked directly for 10 min at 80°C without dilution, cooled in an ice water bath, and then incubated at 43°C for 10 days before plating as described previously. The water used for cooling TRFP and other flavors was also assayed for the presence of the organisms. Samples were filtered through a 0.45 μm membrane filter, 8 mL plated onto K Agar and incubated at 43°C for 4 days, after which they were evaluated for TAB. Swabs taken of equipment at different points along the processing line were assessed for the presence of TAB. Swabs were taken from a 10 × 10 mm^2 area, unless taken directly from specific parts of the processing equipment (e.g., filler tips) where sufficient surface area was not available to use the template. In these cases, the entire equipment part was swabbed. They were moistened with 0.1% peptone water prior to use and the peptone water used to recover the organisms which were plated as previously described. In all cases, any plates on which organisms were found were evaluated for the presence of guaiacol, as described previously.

Identifying the causative agent

In assessing the possible causes of the spoilage, it was determined, based on extensive sensory analyses that it had to be of a microbiological origin. These microorganisms had to be able to either survive the thermal process being applied or they had to be getting back into the product after heating. Further, they had to be able to survive and grow in a very hostile, acid environment and to metabolize materials available in the juice drinks. They also had to be of a type that did not produce gas on spoiling the product as none of the problematic beverages showed signs of gas formation at any point throughout the process. This pointed to spore forming bacteria that could tolerate acid environments, the most likely being members of the group of organisms causing *flat sour*–type spoilage. Among the major flat sour spoilage organisms are *Bacillus stearothermophilus* and *Bacillus coagulans* (Knight and Gordon, 2004), both of which are aciduric, facultatively anaerobic, thermoduric spore-formers. *Bacillus megaterium*, an aerobe, could also to be considered as a potential causative agent for flat sour spoilage of acidic beverages (Tianli et al., 2014). Research had also indicated that the kind of spoilage being experienced could also be caused by an emerging group of organisms known as TAB. Consequently, investigations into the causative organisms were extended to also cover this group of bacteria. LAB, yeasts, and molds were eliminated as possible causes because of their inability to survive the pasteurization process applied. To determine the cause of the problems being experienced, a sufficiently comprehensive approach had to be developed that would ensure the recovery and enumeration of all possible organisms that could cause the problem. Such an approach was adopted by TSL and is outlined in subsequent sections.

The problem with the product had been reported as a sporadic finding in Tropical Rhythms Fruit Punch (TRFP), the first and major flavor of the new line of products. The reports suggested that the organisms responsible consistently produced an off-odor and off-flavor in the product that was medicinal in nature. Further research indicated that this finding was typical of the TAB group of organisms and of *Alicyclobacillus* in particular. Biochemical differentiation of the organisms recovered was therefore undertaken to differentiate them, after preenrichment as previously described. The biochemical profile expected from *B. coagulans* and *B. stearothermophilus*, two other potential causes of the spoilage, and *Alicyclobacillus* spp. are presented in Table 9.1.

Isolation and enumeration of thermophilic acidophilic bacteria

The TAB isolated from the finished products were enriched to assure ease of recovery, and then enumerated using K Agar as previously described. TSL also developed enrichment methodologies (indicated previously) to facilitate more luxuriant growth of the TAB in susceptible substrates, thereby allowing for easier quantification or at least detection of their presence or that of their metabolites. Consequently, if the TAB were present in a particular finished product or ingredient, it was now possible to definitively differentiate between TAB-free and contaminated ingredients and finished products.

Table 9.1 Features distinguishing *Alicyclobacillus, B. coagulans,* **and** *B. stearothermophilus*

Feature	*Alicyclobacillus* spp.	*B. coagulans*	*B. stearothermophilus*
Gram stain	+ve/variable	+ve	+ve
Catalase	+ve[a]	+ve	+ve
Motility	+ve	+ve	+ve
Indole	−ve	ND	ND
VP reaction	−ve	+ve[v]	+ve
Acid from D-mannitol	+ve	ND	ND
Acid from D-glucose	V	ND	ND
Acid from L-arabinose	+ve[a]	ND	ND
Acid from D-xylose	+ve[a]	ND	ND
Acid form D-trehalose	V	ND	ND
Citrate	+ve	ND	ND
Reduction of nitrate	−ve	ND	ND
Growth in 5% NaCl	−ve	+ve	+ve
Growth in 7% NaCl	−ve	−ve	−ve
Aerobe	+ve	+ve	+ve
Facultative anaerobe	+ve[b]	+ve	+ve
Resistance to lysozyme (0.001%)	+ve[a]	−ve	−ve
Growth in anaerobic agar	−ve	+ve	+ve
Growth @50°C	+ve	+ve	+ve
Growth @65°C	+ve	−ve	+ve

sp, Species (all species represented here); V, the results are variable for glucose and trehalose; v, this property may be variable.
[a]Most strains are positive.
[b]Some strains, *A. acidoterrestris* is an obligate aerobe.

The results of a range of microbiological assays done on the isolates after subjecting them to various conditions including varying pH, growth temperatures, and the presence or absence of selected substrates are summarized as follows:

- There was a significant increase in the numbers of acidophiles recovered from the samples after a 1-day enrichment at 43°C.
- Although growth was obtained, the colonies were found to be smaller when cultured at 35°C (Table 9.2).
- No growth was observed at 55°C and above.
- The best (most abundant) growth was obtained at 43°C.

The findings further indicated that the optimal temperature for growth of the organisms in question was 43°C. They also showed that, in some cases, there were differences in the numbers of vegetative cells and spores present in the same sample of product. This, among other factors, indicated the need to separate *the prevention/ inhibition of spore outgrowth* from *the elimination/inhibition of any vegetative TAB cells that may be present in the product.*

Table 9.2 Characteristics of organisms isolated from Tropical Rhythms Fruit Punch

Characteristic[a]	A. acidoterrestris	Isolate A1 from juice coded 1184	Isolate A2 from juice coded 1184	Isolate A3 from juice coded 1181	Isolate A4 from juice coded 1181
Gram stain	+ve/variable	Spores stain GP, bacilli	GP bacilli with spores	Spores stain GP, bacilli	Spores stain GP, bacilli
Catalase	+ve[b]	−ve	−ve	−ve	−ve
Motility	+ve	+ve	+ve	+ve	+ve
Indole	−ve	−ve	−ve	−ve	−ve
VP reaction	−ve	−ve	−ve	−ve	−ve
Acid from D-mannitol	+ve	−ve	−ve	−ve	−ve
Acid from D-glucose	V	−ve	−ve	−ve	−ve
Acid from L-arabinose	+ve[b]	−ve	−ve	−ve	−ve
Acid form D-trehalose	V	ND	ND	ND	ND
Citrate	+ve	ND	ND	ND	ND
Growth in 5% NaCl	−ve	−ve	−ve	−ve	−ve
Aerobic growth	+ve	+ve	+ve	+ve	+ve
Facultative anaerobe	+ve[c]	−ve	−ve	−ve	−ve
Hydrolyze starch	+ve	ND	ND	ND	ND
Resistance to lysozyme (0.001%)	+ve[b]	ND	ND	ND	ND
Growth in anaerobic agar	−ve	−ve	−ve	−ve	−ve
Growth @50°C	V	ND	ND	ND	ND
Growth @65°C	V	−ve	−ve	−ve	−ve
Growth @55°C	V	−ve	−ve	−ve	−ve
Growth @35°C	+ve	+ve	+ve	+ve	+ve

ND, Not done; V, the results for this assay of this property are variable.

[a] The reaction for acid production from sugars was found to be very problematic. Caution is required in interpreting these results.

[b] Most strains are positive.

[c] Some strains.

Table 9.3 **The effect on TAB growth on the pH of K-medium supplemented with mannitol and D-glucose**

Isolate	Isolate A1	Isolate A2	Isolate A3	Isolate A4
pH of control mannitol	4.43	4.43	4.43	4.43
pH after 5 days growth[a]	4.72	4.59	4.64	4.64
pH of control D-glucose	4.25	4.25	4.25	4.25
pH after 5 days growth	4.46	4.47	4.50	4.50

[a] K-medium was used as the basal growth medium.

Identification of TAB

Growth, biochemical and other characteristics were used to identify the TAB isolated from spoiled TRFP, differentiate them from other similar bacilli and assign them to appropriate genera and, if possible, species (Table 9.1). The details of the assessments undertaken and the results derived are presented in Table 9.2. In addition, further evaluation of the ability of the isolates to acidify selected media and the effect of their growth on these media were used to try to differentiate between different species of TAB. These data are presented in Table 9.3. The comparison of the biochemical profile of the TRFP isolates with that which would be expected for the *Bacillus* spp. indicated that they were not *B. coagulans* or *B. stearothermophilus*. The key differentiating characteristics were the fact that they were obligate aerobes (they did not grow when incubated anaerobically), were resistant to lysozyme and were Voges-Proskauer negative. These findings were typical of *A. acidoterrestris* but not of the *Bacillus* spp. targeted. These findings led to further assessment of the isolates, all of which were shown to be alicyclobacilli, though, in some cases, atypical *A. acidoterrestris* as shown in Table 9.2 where several of the characteristics of the isolates varied from those for classical *A. acidoterrestris*.

Summary of key outputs of the first phase of an approach to a solution

The first phase of the approach to a solution successfully adapted and developed methodologies for isolating and recovering the causative spoilage organisms, thereby allowing for definitive determination of their presence in finished product. This same approach could now also be applied to raw materials and the production environment, facilitating specific identification of the source of the problem and hence, a solution (in the second phase, discussed in subsequent sections). Further, Grace now had a means of effectively screening finished product for the problem, even while a solution was being developed, thereby facilitating release of products already made. Importantly, all of the assessment were applicable in a commercial production setting and could be used immediately to address the current issue.

The results of the characterization of the isolates from the spoiled TRFP indicated that the spoilage organisms were TAB and not other types of spore-forming bacilli, and that they appeared to be atypical *A. acidoterrestris* strains. The biochemical reactions differentiated the organisms from *B. coagulans* and *B. stearothermophilus* (Table 9.1). Their optimal growth temperature also eliminates *A. acidocaldarius* (Table 9.2). This was confirmed by the results of independent assessments of samples sent to Campden & Chorleywood Food and Drink Research Association—CCFDRA (now Campden BRI) in the United Kingdom, including phylogenetic analyses using a Dupont Ribo-Printer that definitively placed the isolates as atypical *A. acidoterrestris*. This meant that the firm now had a definitive identification on the causative agent and could now employ specific preventative measures from among those available at the time, as advised by the TSL team of technical advisors. An interim fix was applied as follows:

• Screening of all incoming raw materials for *Alicyclobacillus*.
• Ensuring the effectiveness of the thermal process applied in destroying all vegetative cells of all organisms possibly present while a detailed assessment of the efficacy against *Alicyclobacillus* was done.
• Insertion of a check to ensure rapid and effective cooling of the product to prevent the outgrowth of any surviving spores.
• Positive release of all batches of product based on results indicating that they were free of *Alicyclobacillus* immediately and within 10 days postproduction, using the enrichment approach described previously.

Having put these interim measures in place to facilitate continued supply to the market, the technical team then moved to focus on fully characterizing the problem, identifying its source and implementing mitigating measures. In the meanwhile, work was also undertaken to assess the effectiveness of using nisin as an additional barrier to growth of *Alicyclobacillus*, given its identification as an effective agent in preventing spore outgrowth among TAB (Komitopoulou et al., 1999; Yamazaki et al., 2000). Phase two of the process of implementing an effective solution is described as follows.

Phase II: identification of sources of contamination and the development and implementation of preventative measures

Detection of TAB in sugar, fruit purees, concentrates, and other flavors of Tropical Rhythms

All of the fruit purees and concentrates used in making of the products, as well as the sugar syrup (60 Brix) and all other flavors of Tropical Rhythms besides TRFP were assayed for the presence of TAB using a similar approach to that used to isolate and enumerate the TAB from the spoilt TRFP (described previously). For the sugar syrup, the samples were heat shocked for 10 min at 80°C before plating. After preenrichment, purees were also evaluated organoleptically for the odor of guaiacol or related off-odors.

The results of these analyses on drinks and purees are present in Tables 9.4 and 9.5, respectively. The evaluation of the concentrates and the other flavors of drinks did not find any alicyclobacilli or TAB in any of the samples tested at this stage. Also, no TAB were recovered from the sugar syrup. This is important since sugar stock solutions have previously been identified as a source of *Alicyclobacillus* (Silva et al., 1999) and are used in all of the beverages made in the plant.

Table 9.4 **Results of** *Alicyclobacillus* **detection on the fruit concentrates**

Concentrate (origin)	Brix before heat shocking	pH after dilution	Total aciduric bacteria count in 8 mL	Guaiacol odor detected
Passion fruit concentrate (Ecuador)	13.8	2.96	0	None
Lime juice concentrate (USA)	14.2	2.35	0	None
Orange juice concentrate (Belize)	13.7	3.83	0	None
Pear juice concentrate (USA)	12.0	3.93	0	None
Apricot concentrate (USA)	11	3.27	0	None
Red papaya concentrate (Mexico)	8	3.24	0	None
Mango concentrate (Columbia)	13.6	3.88	0	None
Pineapple concentrate (Thailand)	14.1	3.85	0	None
Pink guava concentrate (Ecuador)	9.2	3.84	0	None

Table 9.5 *Alicyclobacillus* **count in other flavors of Tropical Rhythms Fruit Drinks**

Product	Total acidophilic bacteria count per mL
Mango Carrot (21354, BB May 2003)	0
Passion Carrot (20812, BB March 2003)	0
Guava Carrot (20993, BB April 2003)	0
Pine Ginger (21362, BB May 2003)	0
Sorrel Ginger (20944, BB April 2003)	0

Note: These samples were incubated at 43°C for 4 days prior to being tested.

While no TAB were found in any of the other flavors of drinks assayed, it was necessary to confirm whether or not this was because of an innately different and possibly inhibitory composition of the these products or other factors that were not immediately evident. They were all produced in the same plant using the same equipment and process and, in several cases, shared constituents with Tropical Rhythms Fruit Punch (TRFP). Follow-up studies on the growth of the TAB inoculated into each drink were therefore undertaken, and these indicated that *Alicyclobacillus* grew well in all of the other flavors of Tropical Rhythms except for Tropical Rhythms Pine-Ginger and Tropical Rhythms Sorrell. Both of these contained a ginger emulsion that was not used in TRFP. Ginger has been reported to have at least bacteriostatic, if not bactericidal effect on a range of organisms, including Gram-positive bacteria (Tan and Vanitha, 2004), which may have been the reason for the inability of the TAB to grow in both flavors. This is an area in which formulation may provide some opportunities for creating more robust acidified drink products and merits further research. The growth of *Alicyclobacillus* in other Tropical Rhythms products, nevertheless showed that they were also at risk of spoilage due to contamination with the organism if they become infected during processing.

Metabolic by-product detection

A methodology was developed for identifying guaiacol in products or ingredients that might have had even trace levels of the contaminant present. This was important as a further means of ensuring that product to be released for sale were free of taint caused by the presence or growth of TAB. The method involved a preenrichment process, followed by the determination of guaiacol by high performance liquid chromatography (HPLC) as described by Lee and Nagy (1990), adapted for using an Agilent 1100 series HPLC with a Zorbax 300 Extend C-18 column and fluorescence detector, supported by a Chemstation-based data management system. (Fig. 9.7). To ensure the sensitivity of the methodology, standard solutions and a range of product spiked with guaiacol were also assayed, with every assay including positive and negative controls. The application of the HPLC method for detection of guaiacol down to the level of less than 100 parts per billion (ppb) allowed for early detection and interdiction of tainted

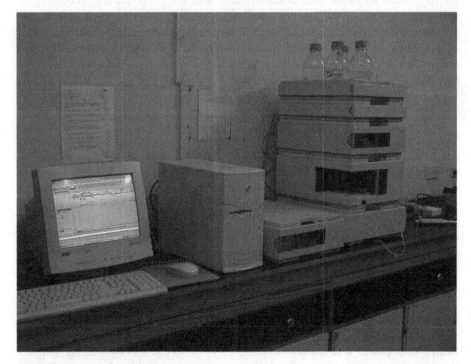

Figure 9.7 Agilent 1100 series high performance liquid chromatograph used for guaiacol determination.
Source: André Gordon, 2002

product and its removal from the commercial supply chain, thereby further reducing the risk of unfit product getting to market.

In developing and deploying analytical regimes as part of a quality assurance program for the acidified juice drinks, another concern was that the precision of the testing being done was subject to the vagaries of batch-by-batch variation, as is the case for all noncontinuous analyses. This meant that redundancies needed to be built in whereby the risk of missing contaminated product due to sampling errors and batch by batch variation were reduced. This magnified the importance of having another method of determining whether the product was free of the presence and activities of the TAB through metabolic by-product (taint) determination. The capability to detect *Alicyclobacillus* and other TAB, if present, in all relevant raw materials and other inputs was important because it greatly assisted in solving the problem by ensuring the use of TAB-free raw materials and other inputs. The ability to enrich for and detect halophenol (guaiacol) taint was also critical as it provided another tool to protect the integrity of the product being sold and reduce the risk of tainted product getting to market.

In controlling or eliminating the risk of *Alicyclobacillus*/TAB in acidified fruit beverages, testing of all raw materials prior to use is important as, done effectively, it can assure freedom from TAB contamination of finished product if steps are then taken to ensure that contamination does not occur during processing. A major finding, therefore, that impacted the implementation of a solution for the problem being experienced with Tropical Rhythms fruit juice drinks was that *great care must be taken to ensure that no raw material introduced* Alicyclobacillus *into the production system and that all needed to be certified as* Alicyclobacillus-*free before use.*

Identification of sources of contamination in the production plant

The fact that the raw materials had been shown to be free of contamination and were not the source of the atypical *Alicyclobacillus* that was causing the problem meant that there was some other source of contamination and that this had to be located to effect a sustainable solution to the problem. Consequently, other aspects of the production system were examined to identify the points at which contamination of the products was occurring. This included the water used in the manufacture of the product which, in the past, has been found to be a source of *Alicyclobacillus* contamination, cooling water, and the equipment used in processing.

Detection of TAB in cooling water

The water used for cooling in the cooling tunnel (Fig. 9.6), as well as an input into the product as a raw material was another potential major source of *Alicyclobacillus* as previously noted. Cooling and process water, both from the same reverse osmosis-treated source, was assayed for the presence of the organisms by filtering samples through a 0.45-μm membrane filter, plating onto K Agar and incubating at 43°C for 4 days, after which they were evaluated for *Alicyclobacillus*. All samples evaluated over time were found to be free of the organism.

Table 9.6 Evaluation of equipment swabs for TAB

Site	Total count/mL
Premix Stock Tank	2500
Filler Tip 19	0
Batching Tank 1106	0
Interior of Mixing Tank #2	0
Concentrate Blending Tank	0
Transfer Pipe Leading from Sugar Tank	0
Filler Tip 3	0
Filler Tip 22	2500

Identification of potential contamination from equipment

To identify any possible sites of contamination throughout the production system, swabs were taken of equipment at different points along the processing line and assessed for the presence of TAB, both before and after sanitation of the production line. The results (Table 9.6) showed that TAB were found in the Premix Stock Tank and on Filler Tip 22 (Fig. 9.6). This clearly indicated that a persistent infection of TAB had been built up at various points on the processing line over time and that these were not being effectively addressed by the existing sanitation regime. These findings were repeated in subsequent assessment, confirmed the persistence of higher than acceptable levels of microorganisms in these two areas during routine postsanitation swab checks (for total bacterial numbers—aerobic plate counts). These findings indicated that:

- there was a systemic weakness in the cleaning and sanitation regime being applied to these two areas of the production line at the time;
- this being the case, TAB were able to establish a persistent infection in the Premix Tank and the Filler Tips; and
- TAB would also be able to establish sites from which on-going contamination of the beverage would arise unless the efficacy of the cleaning regime and the effectiveness of the combination of cleaning agents and sanitizers being used against these specific organisms were significantly improved.

In summary then, the finding of *Alicyclobacillus* on a filler tip and in the Pre-mix Stock Tank particularly at a time when no TRFP or other flavors had been run for some time indicated that a persistent infection had been established in selected areas of the production line, likely as a result of an ineffective sanitation program. This indicated the need for greater stringency and care in the cleaning and sanitation of the production line in the plant.

This definitive aspect of the assessment for the source of *Alicyclobacillus* in the system, while not identifying the original source, confirmed why there was an ongoing and sporadic contamination of the juice drinks. These findings also suggested that other areas of the production line that could harbor bacteria may also have been or became secondary sources for *Alicyclobacillus* and needed to be routinely checked for these organisms as part of an on-going swab (sanitation efficacy) checks. They further showed that not only would the problem of ongoing contamination persist, but that it would be sporadic as it depended on which bottle of product was filled at a contaminated filler tip and which was not, making detection of the problem difficult. Of greatest concern is the fact that a filler tip was found to be positive because contamination at this point, which was after the thermal process step, was likely to negatively affect the shelf life and acceptability of the product since only cooling remains as a possible control measure after this stage.

Ensuring the effectiveness of the sanitation regime

The identification of reservoirs of *Alicyclobacillus* along the production line led to an evaluation of the sanitation program. It was realized that the program needed to be modified to get it working in such a manner that effective cleaning and sanitation routinely occurred, particularly postprocessing. This was addressed through a detailed review of the cleaning protocols and the sanitizers being used and the recommendation and institution of a more rigorous and effective cleaning and sanitation regime. This regime was based on rotating the sanitizers used and introducing the use of quaternary ammonium sanitizers which are more effective against Gram-positive organisms than hypochlorite alone, and are very effective against *Alicyclobacillus* (dos Anjos et al., 2013). This program had to be validated as effective and this effectiveness monitored by the ongoing monitoring program that measured and documented the efficacy of the cleaning and sanitation regime now incorporating a specific check for TAB and *Alicyclobacillus*.

Evaluation of the effectiveness and impact of other aspects of the process

The efficacy of the thermal process and other aspects of the process applied

Tropical Rhythms Fruit Punch (TRFP), the flagship product and the one with which the problem was first and primarily observed, as well as the other flavors, were manufactured using the process shown in Fig. 9.5, a key step in which was the thermal processing of the fruit drink. This process involved heating the product in a heat exchanger at 92°C for 30 s, a process which was sufficient to meet the FDA's 5-Log reduction target and eliminate vegetative cells of the target pathogens, as well as other known spoilage organisms (thermoduric yeasts and molds) at that time. The findings of TABs as contaminants of the TRFP postprocessing, of the fact that they could survive and grow in other variants of the product and that ingredients and the processing line itself could possibly be a source of future contamination, however, meant that the process had to be rigorously examined if it was to be modified to be more effective. Research done by TSL at the time had shown that while the process would kill the vegetative cells of *A. acidoterrestris*, it would not be able to destroy its spores which have a D_{91} of 10–20 min and z-value of 5–22°C (Pontius et al., 1998; Silva et al., 1999; Splittstoesser et al., 1994). This presented a dilemma as the option of using a process that would destroy alicyclobacilli spores was not available as it would also completely render the product unpalatable.

A review and understanding of all of the facets that influence the efficacy of a thermal process for organisms which often cannot be controlled by various time/temperature combinations alone provided the solution. As *A. acidoterrestris* is an obligate aerobe, it requires the presence of air (oxygen) for it to grow in the fruit drinks. This means that wherever organisms or guaiacol was detected, those bottles of product had not developed a hermetic seal[f] and a sufficient postprocess vacuum to exclude the presence of air in the headspace[g] of the bottled product. This meant that if sufficient focus was placed on ensuring the development of a good vacuum and a hermetic seal, even though the TAB would survive the process, their growth would be impaired, if at all possible in the bottles of juice drinks because of the anaerobic environment in the container. Among the ways of ensuring a good vacuum and hermetic seal in a container is the use of rapid cooling down to room temperature or below. This aspect of the process therefore became a target for control.

It also became evident that the cause of the problems with the product was more likely to be the spores of *Alicyclobacillus* present, rather than the vegetative cells, most or all of which would have been killed by the thermal process applied. These spores would require the appropriate conditions under which to grow out and begin to cause spoilage, such as the favorable temperatures afforded by slow cooling. Slow cooling could result in the product spending an extended period in the temperature range of 40–50°C which is the temperature range in which *A. acidoterrestris* and most other TAB thrive (Tianli et al., 2014). Consequently, it was reasoned that if the efficacy of the cooling process could be assured such that the product was rapidly cooled and did not spend an extended period of time in the danger zone, then not only would the organisms be denied the time/temperature combination required for them to grow, but a good vacuum and hermetic seal would be assured.

Implementation of solutions to the thermal and other processing challenges

Based on the foregoing findings and assessments, Technological Solutions Limited (TSL), the technical advisors to the firm recommended and Grace implemented the following solutions:

1. Grace ensured that no processing was started until the lines had been cleared as TAB-free by the test results of the postsanitation swab checks (this was after the shut down).
2. The company should ensure that the thermal process being used was developed or validated by a Thermal Process Authority (TPA) and was effectively monitored to ensure that the proper time/temperature combination was always administered and recorded. This was done by TSL's principal consultant,[h] a recognized TPA.
3. The cooling process was to be engineered and managed so as to be able to deliver the rapid cooling of the product as required to avoid spore outgrowth.
4. Cooling water was to be checked to ensure that it had a minimum residual chlorine, was therefore of good quality and would not be a potential source of contamination.
5. A dud detector was to be installed to check each bottle prior to packaging to ensure that it had an appropriate vacuum and to automatically reject those that did not.

Modification of the production process to prevent the recurrence of TAB-mediated spoilage

The information gathered during the process of assessing the nature and causes of the spoilage problems being experienced with Tropical Rhythms Fruit Punch allowed recommendations to be made and systems to be developed and implemented that have successfully prevented a recurrence of the problem. The results of the series of studies that advised the solutions developed identified the areas that required focused attention to prevent contamination of the product and, if present, eliminate the TAB or prevent their outgrowth. This included the development of an effective and efficient sanitation program and methods that allowed for accurate prediction of the growth of *Alicyclobacillus* in Tropical Rhythms products, regardless of blend. In addition, the characteristics of the *Alicyclobacillus* isolates, as presented in Tables 9.1 and 9.2, allowed for adequate control of the organism even if it was present, by providing hostile conditions for TAB growth, thereby inhibiting them. These included an effective kill of vegetative cells by the thermal process and management of the cooling process and postprocess conditions to prevent outgrowth of any surviving spores, including ensuing that all bottles of product to be sold presented an anaerobic environment as indicated by a strong vacuum (the dud detector). These, together with the monitoring of raw materials, checks on the finished products and an effective sanitation program not only significantly reduced the economic risk of production of TRFP and the other beverages, to the barest minimum, they have eliminated the problem for the last 14 years over which there has not been a single recurrence of the spoilage problem.

The specific process modifications (excluding those made to the sanitation program) are shown in Fig. 9.8. The key control steps are highlighted in gray and dark gray.

Case study summary: preventing economic losses for Tropical Rhythms Fruit Juice Drinks due to Alicyclobacillus

TSL has, through research, observation, and discussions with the Grace team, developed and presented recommendations for adjustment to the process and process control procedures to prevent spoilage of the firm's fruit juice drink beverages. This also covered any potential compromising of product quality during storage or at any stage prior to consumption. This was augmented by supporting work done at laboratories in Campden BRI (in the United Kingdom). Overall, the implementation of these recommendation, captured in a revision to the overall process as well as the quality assurance program applied to include a specific focus on alicyclobacilli, was effective in preventing a recurrence of the problem.

The approach, what was done and the findings can be summarized as follows:

- The sporadic spoilage being experienced with the Tropical Rhythms Fruit Punch (TRFP) juice drink was found to be caused by TAB.
- The characterization of the isolates from the spoiled TRFP identified the TAB as atypical *A. acidoterrestris* strains.
- Methods were developed that facilitated effective screening of all raw materials for the presence of, and/or susceptibility to spoilage by *Alicyclobacillus*. A monitoring program for all raw materials prior to acceptance for use was immediately implemented, consequent on this finding. This was augmented by information requested from suppliers of juice concentrates and other inputs asserting the *Alicyclobacillus*-free status of their product.

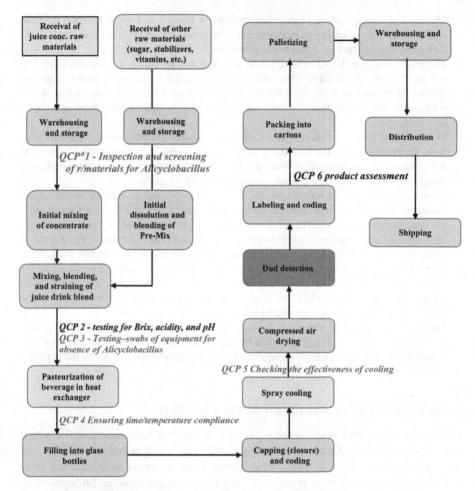

Figure 9.8 Modified commercial process for Grace Tropical Rhythms Juice Drinks postintervention. [a]QCP—the process of testing at this QCP (Quality Control Point) was completely revised to ensure that detailed screening of all inputs was done for *Alicyclobacillus* and other TABs. This was despite the mandatory requirement for certificates of guarantee from the suppliers that their raw materials were TAB-free.

- The assessments were routinely done and in advance of the use of all batches of raw materials.
- It was found that, while not yet a problem, the TAB could grow in all other Tropical Rhythms products, the exception being Tropical Rhythms Sorrel and Pine Ginger which appeared to have an inhibitory effect on them. Nevertheless, careful handling of all of the products was implemented, in keeping with what was being done for TRFP.
- The *A. acidoterrestris* strains found and which were unable to grow in the Sorrell and Pine Ginger flavors may possibly have been inhibited by the ginger emulsion used in both as there were no other major differences between these and other formulations. *This finding still needs to be confirmed.*

- The TAB (*A. acidoterrestris*) were also found to be present in Tomato Ketchup (and the tomato paste from which it was made), a product that was also made in the same plant, indicating the need for care to eliminate the potential risk of cross contamination.
- *Alicyclobacillus* was found to have set up a persistent infection in at least one Filler Tip and the Pre-mix Stock Tank. This indicated the need for a rigorous review and an upgrading of the sanitation program for the fruit juice drink line which was done as it was evident that effective sanitation alone could significantly reduce the risk of spoilage due to *Alicyclobacillus*.
- The control of filling as well as postcooling temperatures is critical to controlling the likelihood of spoilage due to the survival of vegetative cells and/or the outgrowth of spores of *Alicyclobacillus*, respectively. A focus on these, as well as the thermal process itself was used to significantly improve control and reduce the risk of spoilage.
- As the causative organism for the spoilage, *A. acidoterrestris*, was an obligate aerobe, ensuring that an anaerobic environment was created and maintained in each of the bottles of product manufactured was critical to preventing its outgrowth in the product, if present, should cells survive the thermal process. This was assured through the inclusion of a dud detector in line that would detect and remove any bottles of product without an effective vacuum.
- Methodologies to quickly screen finished product for spoilage, incipient spoilage, or the potential for spoilage was developed and tailored for use in a processing plant. This allowed for a restart of the sales of TRFP and confidence in the on-going production and sale of the product range, once the appropriate assessment was done on each batch.
- The program of implementation of the agreed approaches was subsequently reviewed at least annually by the firm to ensure that the latest information available on the issue was considered and, where relevant, incorporated into prophylactic approaches to the potential problem.

Supplemental information on Alicyclobacillus

Subsequent to the work done on the problem with the fruit beverages, parts of which were reported first by Knight and Gordon (2004) and which is fully documented in this case study for the first time, other studies have presented additional information and advances have been made in the detection, enumeration, and mitigation of *Alicyclobacillus*. These are covered in review articles by Ciuffreda et al. (2015) and Tianli et al. (2014) and will therefore not be repeated here. It was felt to be useful, however, to highlight some of the more recent information for tropical fruit juices and juice drinks and advances in detection, enumeration, and treatment. This was felt to be particularly germane to those advances related to the isolation and recovery of the organism that would not have been available in the early to mid-2000s when the author was dealing with the problem of *Alicyclobacillus* contamination that was dealt with in this case study. The most recent advances and new approaches to dealing with the recovery and identification of *Alicyclobacillus* spp. are presented in Table 9.7. Also to be taken into consideration are the findings of Oteiza et al. (2015) with respect to the negative effect of natural antimicrobials present in some citrus juices on recovery and enumeration of the organism postenrichment.

While the methods of detection and quantitation as outlined in Table 9.7 are important, of perhaps greater importance is a summary of the various recent approaches, ingredients, and interventions that have reported some success in controlling the

Table 9.7 New and current approaches to recovery, enumeration, and identification of *Alicyclobacillus* spp.

Method	Description	References
Immunomagnetic separation (IS) and real time polymerase chain reaction (RT-PCR)	Immunomagnetic separation was combined with RT-PCR by using two probes. The method is highly selective for *A. acidoterrestris*	Wang et al. (2014)
Double antibody sandwich ELISA (DAS-ELISA)		Li et al. (2014)
Lipase and esterase fingerprints	Cell harvesting and chromatography after incubating juice at 45°C for 24 h.	Cai et al. (2015)
Aptamer-based enrichment 16S rDNA	A preliminary enrichment step taking up to 7 days produced substrate to which a mechanical treatment was applied. DNA was quantified through RT-PCR	Hünniger et al. (2015)
Fourier transform infrared (FTIR) spectroscopy	Fourier transformed infrared spectroscopy (1350–1700 cm^{-1}) was combined with principal component analysis and class analogy, multivariate statistical techniques, to discriminate between *Bacillus* and *Alicyclobacillus* species	Al-Holy et al. (2015)
G-quadruplex colorimetric method	Guaiacol produced by *A. acidoterrestris* grown at 45°C in the presence of vanilla was converted to tetraguaiacol by G-quadruplex DNAzyme. Tetraguaiacol is brown in color and is then read by colorimetry	

Adapted from Tianli Y., Jiangbo, Z., Yahong, Y., 2014. Spoilage by *Alicyclobacillus* bacteria in juice and beverage products: chemical, physical, and combined control methods. Compr. Rev. Food Sci. Food Saf. 13 (5), 771–797 and Ciuffreda, E., Bevilacqua, A., Sinigaglia, M., Corbo, M.R., 2015. *Alicyclobacillus* spp.: new insights on ecology and preserving food quality through new approaches. Microorganisms 3 (4), 625–640.

Alicyclobacillus problem in juices and juice drinks (Ciuffreda et al., 2015). This is presented in subsequent sections.

The measures that have been used or evaluated for use to control *Alicyclobacillus* spp. in acidified beverages fall into three categories: physical measures and approaches: chemical, physical, and combined control methods and combination approaches which use a mixture of both. The physical measures that have been tested for their efficacy against *Alicyclobacillus* include high electric field (HEF), pulsed electric field (PEF), high hydrostatic pressure (HHP), gamma irradiation (GI), pulsed light (PL), ultrasound (US), high pressure homogenization (HPH), microfiltration/ultrafiltration (MF/UF), microwave (MW), ohmic heating (OH) and short wavelength ultraviolet light (UV-C). The natural compounds and other chemicals that have been used include a range of essential oils, bacteriocins, and other chemicals such as lysozyme, silver nanoparticles (Ag^+), neutral electrolyzed water (NEW), fatty acids, ozone, dimethyl carbonate (DMC), supercritical carbon dioxide (SCO_2) and the preservatives sodium benzoate and potassium sorbate (Ciuffreda et al., 2015; Tianli et al., 2014). In addition to these, a wide range of combinations have been tried, the outcomes of which are well reviewed by Tianli et al. (2014) and will not be repeated here.

The physical methods tried have shown varying effectiveness against *Alicyclobacillus*, with their effect being both strain and treatment dependent. Of the methods examined, the cost of many may be prohibitive for the average acidified beverage manufacturer and so those such as HEF, HHP, OH, HPH, PL, US, and GI may not find widespread use in the industry in developing country although some major manufacturers might want to consider them. PEF has good efficacy against spore-forming bacteria and has been successfully demonstrated in fruit juices with minimal effects on freshness characteristics (such as color, pH value of flavor compounds). Again, however, its efficacy in nonclarified tropical juices and juice drinks would need to be verified and the specialized equipment required and the cost of deploying it may limit its applicability.

Of the nonthermal technologies explored previously, MW, MF/UF, and UV-C may be more feasible. MW has been shown to result in 2-Log reductions for one strain of *A. acidoterrestris* when applied at full strength for 5 min. However, its effectiveness depends on the power and time of exposure and may be affected by the type of juice/juice drink and the particular organism. MF/UF technology has been widely used in the industry for some time and does have some efficacy in the removal of bacteria from clarified juices. However, questions remain about the efficacy of filtration in removing *Alicyclobacillus* due to findings that suggest that spores of the organism are able to penetrate the membrane used (Tianli et al., 2014). UV-C radiation is a nonthermal treatment during which no known toxic or significant nontoxic by-products are formed and requires very little energy compared to pasteurization. UV-C light, which is short wavelength UV light (200–280 nm), appears to hold promise as a treatment for clarified juices because of its relative affordability. Its applicability to, and efficacy with nonclarified juices and juice drinks still, however, need to be determined.

More widely studied and, perhaps, more easily applicable are the use of a range of natural compounds and chemicals. Compounds such as nisin [which exhibit bactericidal activity and is the only bacteriocin approved by the FDA as Generally Recognized as Safe (GRAS)], lysozyme (which is strain dependent), essential oils, cinnamaldehyde, citrus extracts, and eucalyptus extracts have been shown to have efficacy against *A. acidoterrestris*. Rosemary has also recently been shown to be effective in controlling this organism in apple juice (Piskernik et al., 2016). In addition, either sodium benzoate or potassium sorbate at 0.05% have been found to be effective against alicyclobacilli in apple juice and achieved successful inhibition of high levels of inocula (Walker and Phillips, 2008). Micronized benzoic acid used at 50 mg/L in orange juice concentrate, exhibited a 1-Log greater reduction in overall spore counts of *A. acidoterrestris* over that obtained using benzoic acid or sodium benzoate (Kawase et al., 2013). Of course, the use of any of these natural additives as antimicrobial agents would have to be verified in each particular juice drink system as their effectiveness may differ, depending on the *Alicyclobacillus* spp. involved and the drink system in which they are to be applied.

Effective sanitation can be achieved by the application of ozone at 1.8–2.8% (Tianli et al., 2014), chlorine, sodium hypochlorite, and hydrogen peroxide/peracetic acid/octanoic acid in commercial products or chlorine dioxide at 40–120 ppm. Alternatively, as done in this case study, effective sanitation can also be achieved using an appropriate quaternary ammonium sanitizer in combination with a chlorine-based sanitizer and testing for effective destruction of *Alicyclobacillus* spores by equipment and environmental swabs (following enrichment).

Summary, conclusions, and lessons learned

There have been many reports in the literature of spoilage of juices and juice drinks by *Alicyclobacillus*, many of these being for traditional beverages such as apple and orange juices. Silva et al. (1999) studied the heat resistance of the organism in a range of juices, including cupuaçu, an exotic fruit extract. Oteiza et al. (2011), in a study of a wide range of tropical and other fruit and vegetable juices in Argentina found that except for kiwi and orange, *Alicyclobacillus* was found in juices from all the evaluated raw materials with the percentage of positive samples ranging from 24% to 100%. The highest percentage of positive samples were strawberry, banana, peach, beetroot, mango, carrot, and plum juices. More recently, Oteiza et al. (2014) studied the survival of *Alicyclobacillus* isolated from flavorings in a beverage containing mango, while there have also been reports on the contamination of passion fruit and pineapple juices, the first report of *Alicyclobacillus* contamination of exotic Brazilian juices (McKnight et al., 2010). Also, Zhang et al. (2013) reported on the contamination of a production line for Kiwi products in China. Nevertheless, besides these, the literature remains deficient in reports about problems with *Alicyclobacillus* with tropical beverages and in developing countries. This highlights the importance of this case study which focused not only on tropical juice drinks and reported on the concentrates from multiple countries, but also detailed the sources of contamination along the processing line.

In dealing with the problem of *Alicyclobacillus* contamination in the production of tropical fruit-based acidified beverages and the associated economic threat in 2002/2003, targeted studies were done that identified the source of sporadic spoilage of the nonclarified fruit juice first as TABs, then as atypical *Alicyclobacillus*. Interventions based on a combination (hurdle) approach were developed to eliminate the organism from the finished products and were found to be effective in addressing the problem. The approach involved effective screening of all raw materials and the production line and environment for *Alicyclobacillus*, the application of an effectively controlled pasteurization regime that killed vegetative cells of the organism, rapid cooling of the product to prevent spore outgrowth, if present, and ensuring an adequate vacuum and therefore an anaerobic environment in the finished product with dud detection. This was done against the background of a sanitation program using a combination of chlorine-based and quaternary ammonium sanitizers known to have high efficacy against Gram-positive organisms, supported by aggressive effectiveness monitoring.

In developing the solutions, methods were developed or refined to identify and enumerate the organisms and their metabolic by-products in juices, identify the source of TABs in the product and within the production system and examine the effect of ingredients, processing conditions, and mitigating interventions on its control. Ribotyping, physicochemical, and biochemical methods were used to differentiate *Alicyclobacillus* from similar spore-formers, to identify the species involved and its characteristics. A quality assurance protocol specifically for early detection of the organism, elimination, and prevention of outgrowth, if present, was also developed and successfully implemented. The combination methods approach used in this case study, without the use of preservatives or any of the other prophylactic methods available today, has proven effective over the last 14 years. It is practical and applicable to any production environment. It, along with any of the other options discussed in this chapter should, together ensure that fruit and fruit-based beverage producers can effectively deal with the problem of *Alicyclobacillus*, should it ever arise, or institute successful preventative program with confidence into their production processes.

Endnotes

[a]Commercial sterility is defined as the condition in which all organisms of public health and economic significance are destroyed or inactivated such that the product is safe and stable for its intended shelf life under normal conditions of storage and distribution.

[b]The NFPA became the Food Products Association (FPA) in 2005 and subsequently merged with the Grocery Manufacturers Association (GMA) in 2007. The organization has now adopted the latter name and is called the Grocery Manufacturers Association (GMA).

[c]Passed into law in November 2012 and enforced in June 2015, the Safe Food for Canadians Act and Regulations require that all foods offered for sale in Canada must be demonstrably safe, with the manufacturer/distributor being responsible to provide this assurance.

[d]The author's firm, Technological Solutions Limited (TSL), was contracted to solve the problem. TSL specializes in, among other things, applying science to process improvements and problem solving in the food industry and is a pioneer in food safety and quality systems implementation.

[e]A device that tested the cap of each bottle to ensure that it was concave, indicating an effective vacuum. Those without an effective vacuum would be removed.

[f]A hermetic seal is one in which no gases, including air, can enter a container through the seal after processing because it is sufficiently formed to be "air-tight." The formal definition of a hermetically sealed container is "An air-tight container designed and intended to protect against the entry of microorganisms".

[g]The headspace is defined as "the area inside the container that is void of product".

[h]Dr. André Gordon is TSL'S Principal Consultant and is a recognized Thermal Process Authority by the local and regional regulators in Jamaica and the Caribbean, as well as by the FDA with whom he has filed numerous processes over the last 20 years.

Conclusions and lessons learned: steps for successful Food Safety and Quality System (FSQS) systems implementation

A. Gordon*, H. Kennedy**
*Technological Solutions Limited, Kingston, Jamaica; **Technological Solutions Limited, Port of Spain, Trinidad and Tobago

Chapter Outline

Conclusions and lessons learned

The importance of food safety and quality systems

Throughout this volume, the focus has been on understanding the importance and impact of food safety and quality systems (FSQS) on the production and export (mainly) of food products, many of them traditional, from developing countries to third world country markets. It should be apparent that regardless of location, the regulations, standards, certification, and other market requirements have become, or are rapidly becoming, the same. In this regard, with the primary focus or regulators being on the safety of the food and buyers wanting both safety and quality, a hazard analysis critical control point (HACCP)–based food safety and quality system (FSQS), supported by a comprehensive set of prerequisite programs (PRPs), is the best route for meeting these requirements. Such systems can consistently provide safe, high-quality food and, supported by other programs that conform to the specific requirements of regulations, such as the SFCA or FSMA, will facilitate the market access required to allow the expansion of food-exporting businesses in developing countries. They will also facilitate the diversification of the global food supply by ensuring the delivery of safe, quality food to consumers across the world.

Understanding the role of foodborne illnesses and their impact on regulatory requirements

The importance and role of foodborne illnesses in the global food trade was examined in Chapter 2. Specific outbreaks have led to changes in the way markets address market access issues, as well as changes in the regulatory and market requirements for trade in food. Specific examples of important foodborne illness outbreaks in fruits and vegetables, including those involving the fenugreek sprout *Escherichia coli* O104:H4 outbreak in the European Union in 2011, the 2012 *Listeria* outbreak from Jensen Farms cantaloupes, *Salmonella* outbreaks involving mangos and papaya from Mexico, and the *Cyclospora cayetanensis* outbreak from Guatemalan raspberries and blackberries were discussed. From these cases, it became evident that stringent controls, effective traceability, the ability to quickly and effectively investigate foodborne illness outbreaks, and collaboration across borders is important as the world's food supply becomes more globalized. It should also be apparent that it is critical for production and exporting entities in different jurisdictions to be aware of, and fully understand the regulatory requirements of importing countries, as well as food safety systems required by buyers, such as the Global Food Safety Initiative (GFSI).

It should also be evident that food industry professional and exporters need to have a better understanding of the factors that can cause potential contamination of their products, how to mitigate these risks, and comply with market access requirements. While HACCP-based food safety systems or a FSQS compliant with the GFSI may be acknowledged by regulators as a good step toward complying with their requirements, they are by themselves not sufficient. All countries have specific labeling, ingredient prohibition and product registration and filing regulations that firms must comply with if their goods are to be allowed across borders. This means that food industry professionals involved in global food trade, wherever situated, need to become aware of, and keep abreast with the changing regulatory and market requirements, the constant upgrading of which are likely to continue for the foreseeable future.

Application of food science in improving quality, safety, and market access

Three of the cases explored in this volume demonstrated how the application of food science in the areas of microbiology, food chemistry, food analysis, and food processing, as well as sensory science and other tools can be used to address technical challenges and improve a firm's food safety, quality, and competitiveness. These will collectively improve the market access possibilities for the exporting firm and hence their business prospects. In Chapter 3, in the case of Irish moss (*Gracilaria debilis*) beverages, a systematic approach through a planned series of goal-oriented, food science–based interventions within the reality of a small-developing country firm with limited capabilities was able to address quality challenges and open markets. Like the case of coconut water (Chapter 7), required the application of microbiological studies, shelf-life determination (accelerated shelf-life determination was used for both), and changes to process and systems implementation to provide solutions. These approaches, as well as sophisticated analytical methods, some of which had to be developed in-house at the time, were also used with the quality challenge of *Alicyclobacillus* (Chapter 9), with combined methods (hurdle) technology also being applied in this case as well. The cutting-edge methods employed, such ribotyping, physicochemical, and biochemical methods used to differentiate *Alicyclobacillus* from similar spore-formers, and the investigation of enzymatic degradation, as well as the development of a reliable, practical index of quality and shelf life for coconut water demonstrate the role for sound science in all circumstances. Approaches similar to these can work for firms operating in a developing country environment if they have access to good technical support, even as they will likely face challenges in terms of access to a variety of other resources. In the context of these challenges being addressed more than a decade ago and the significant advancement in information and other technologies, it is expected that FSQS practitioners can take guidance from these examples in addressing current day issues.

Application of appropriate technology for the safety of traditional foods

Traditional food products often do not fall neatly within the systems of classification of importing country markets, particularly where the means by which they are made and their safety is assured are unfamiliar to those markets. This was a major hurdle for Ackee exports to the United States discussed in detail in Gordon (2015a). In Chapters 2, 5, and 6 of this volume, several traditional products that are growing in importance in global trade were explored, as were the means by which their safety was proven to regulatory authorities in North America. These included soy sauce made by different processes, browning, a chutney-type product, jerk seasoning, and canned pasteurized processed cheese. These products were all made safe by the application of a combination of approaches, including ingredient technology, acidity and pH, water activity, and thermal processing, among others. With many of these being nonthermal combined methods (hurdle) technologies the verification of their effectiveness in guaranteeing safety to the satisfaction of regulators is often a challenge for producers. In the case studies covered, the approaches to document the safety of these products made using combined methods (hurdle) technology was demonstrated with specific reference in all cases being made to the applicable regulation under which they fell. In the case of browning and soy sauce, the chemistry of the process formed a part of the submission along with detailed process descriptions indicating where the controls were applied. For jerk seasoning and the canned pasteurized processed cheese, challenges studies with selected pathogens or their proxies, were used. With the processed cheese, shelf-life studies and detailed thermal process analysis, in addition to verifying compliance with the Tanaka principles were also important in demonstrating regulatory compliance. Other foods that fall within these categories should be able to use similar approaches along with the much clearer guidelines that now exist and the guidance provided in this book to demonstrate compliance, if required.

Ensuring compliance with requirements for horticultural-based food imports to the United States

The imposition of import alerts on a canned vegetable item exported to the United States for foreign (extraneous) matter and pesticide residues was examined in Chapter 8. It was demonstrated that to have the prohibition removed, the technical team had to clarify the regulations that applied at the time (1993/94) for both pesticides and insect residues [extraneous matter (XM)]. It was then required to specifically agree with the regulators what limits would be applicable to the crop that was not grown in the United States and therefore it was not evident where they were addressed by US regulations, at least not initially. The technical team also had to undertake detailed studies to identify best practices to reduce the insect burden on the crop, ensure that appropriate postharvest intervals were followed, and implement effective controls in the processing to eliminate the XM from the canned product. The nature of the studies required meant that no exports took place for 13 months until all requirements were fulfilled. The important lessons from this case study are that it is critical to establish a good regulatory interface to understand and agree the import requirements that exporters are trying to meet and also that the MRLs and the DALs for each specific crop must be understood and targeted by the firms' FSQS program. The second case study showed that food industry value chains have a natural tendency to drift from agreed practices over time, resulting in noncompliance with agreed standards, in this case pesticide residues and MRLs. The lesson here was that FSQS specialists need to ensure that systems are put in place to ensure the maintenance of compliance to market entry requirements over time, inclusive of scheduled audits and analyses, where required.

A systematic approach to the implementation of a world-class FSQS in a developing country environment

The process of the implementation of a good manufacturing practices (GMP)– and HACCP-based food safety system was described in detail for Mountain Top Springs Limited (MTSL) in Chapter 4. Driven by the implementation GMPs first, along with other PRPs, this case showed how transformation of a firm's operations, its staff competencies, access to markets and, ultimately, the sustainability and growth of its business is possible in developing country environments. What is particularly enlightening about this case is the comprehensive nature of the approach that was taken, including redoing the plant layout and design, transformation of the human resources of the facility through effective training, full PRP implementation to world-class standards, and institution of an analytical program despite the remote location of the plant. It showed that even in challenging circumstances in a developing country environment, successful implementation of FSQS is possible and can achieve excellent results.

The commitment to changing the culture of the firm by management through an effective, comprehensive program of training and human resource development resulted in the sustainable transformation that was sought. It showed that, if empowered by a management team that trusts and supports them, all staff members can learn and implement the required basic food safety systems and practices, freeing management to focus on growing the business. The case study further confirmed the strong assertions of Yiannas (2008) that changing the culture of an organization is as critical to effective FSQS implementation as all other technical areas of the process. The implementation team should be peopled not just by management and supervisory personnel, but also by some of the natural leaders within the operations who must take time to identify, win over, and nurture. These natural leaders, more than anything else, will provide the impetus to constantly drive the implementation process, as they have the trust and respect of their peers.

Other important lessons learned were that a proper layout and flow, coupled with effective, efficient inventory management, and warehousing and storage will significantly improve the efficiency of the operations, as well as quality and safety. Overall, it was shown that a properly designed and planned implementation program that is target based, milestone driven, and effectively communicated is among the surest ways to have successful, timely implementation. Finally, besides the business benefits that are accrued (and will accrue) to firms, it was shown success is likely if a structured, methodical approach to implementation, which is sensitive to cultural realities and applies creative but globally acceptable solutions to challenges encountered during implementation, is followed.

Summary of the steps for successful implementation of a food safety and quality system

All of the cases considered in this volume and summarized earlier, as well as others have advised the development of an approach to systems implementation that has the greatest likelihood of success in a timely manner. There are several steps involved in implementation of an effective and sustainable food safety and quality system that can be generalized. Several of these are often done at the same time and there may be overlaps, depending on how the firm defines its activities or apportions responsibilities. The steps outlined further should be employed if a comprehensive, effective, and compliant FSQS is to be implemented.

FSQS system implementation can be encapsulated in the following steps:

1. Establish clearly defined strategic food safety/quality objectives.
2. Define the scope of the FSQS.
3. Develop a food safety/quality policy.
4. Commit to compliance (and certification, as evidence of compliance).
5. Develop a FSQS communication plan.
6. Explore various food safety/quality standards, pertinent legislation, and customers' requirements.
7. Select an appropriate food safety/quality standard/scheme.[a]

8. Establish a food safety/quality team.
9. Appoint a food safety/quality team leader.
10. Provide the required resources to the food safety/quality team.[b]
11. Conduct an official launch of the FSQS.
12. Provide appropriate training for the food safety/quality team.
13. Create food safety/quality awareness across the company.
14. Conduct a gap analysis to benchmark the company's current status and identify gaps.[c]
15. Evaluate the elements of the gap analysis and ensure full understanding of the findings, their implications, and their importance by the ownership/management of the company.
16. Establish an approach to the development of a comprehensive FSQS.
17. Develop a project plan for development (documentation of policies, procedures, and work instructions) and implementation of the FSQS.
18. Approve the project plan.
19. Execute the project plan (development and implementation of policies and procedures and work instructions).
20. Develop an internal auditing capability.
21. Periodically communicate the status of development and implementation of the FSQS to key stakeholders.
22. Monitor conformance to and effectiveness of the documented policies, procedures, and work instructions through periodic audits.
23. Conduct periodic scheduled management review meetings.
24. Develop a corrective action plan and implement corrective actions based on the results of the internal audits and management review.
25. Arrange and external audit of FSQS for certification.
26. Conduct a certification audit.
27. Develop a corrective action plan and implement corrective actions based on the results of the certification audit.
28. Maintain certification.
29. Continuous monitoring.
30. Continually improve the FSQS.

A step-by-step guide to the implementation of a food safety and quality system

A step-by-step guide

While there are many approaches to a successful implementation of a sustainable food safety and quality system, the authors' collective experience of over 50 years in the field in multiple countries suggest that some approaches are more effective than others. They have developed an overall approach that, while not cast in stone and clearly malleable based on existing circumstances, has lead and will lead to successful implementation in an efficient and effective manner. This is summarized further.

1. Establish clearly defined, strategic food safety/quality objectives

   Before embarking on the development of a FSQS, an organization should establish clearly defined strategic food safety and quality objectives. The senior management team of the organization, in the strategic planning process, should have not only identified specific food safety and quality goals, but also have determined the measures that will be used to determine progress toward and achievement of these objectives. They should also develop strategic initiatives, a time frame, and specific accountability for the development and implementation work. *Critically, all of this must be tied to the attainment of specific business objectives by the firm.*[d]

2. Define the scope of the food safety and quality system

   In the development of the FSQS, the scope must be clearly defined. The organization must be very clear on whether the FSQS will apply to the entire operation, that is, all the organization's processes and products or only to some aspects, whether the senior management team should decide to focus on specified areas of the operations or specific products initially, and whether it is wise to determine the time frame in which the other processes and products would be incorporated into the system. Appropriate documentation should be maintained for any changes in scope over time.

3. Develop a food safety/quality policy

   Having identified the strategic food safety and quality objectives and the scope of the FSQS, it is recommended that the organization's leadership/senior management team should commence the development of a food safety/quality policy. The food safety/quality policy is essentially a documented expression of commitment to the development, implementation, maintenance, and continual improvement of the FSQS by the organization's leadership and management teams. The policy should be written in language that can be easily understood by all stakeholders and should be shared with all stakeholders. While it may not be necessary for employees of the organization to be able to repeat the food safety/quality policy verbatim, they should have a clear understanding of the contents and the role that they are required to play to ensure its implementation and maintenance.

4. Commit to compliance and certification

   The organization's senior management team should commit ensuring full compliance of the system from the beginning and to the certification of the FSQS very early in the process. Compliance is critical and can be confirmed by a noncertification (e.g., precertification) audit that covers all of the same elements as would typically be required of a certification audit. Compliance is what determines the acceptability of the food products being packaged and sold by the firm. Formal certification of the FSQS confirms that the organization's system meets the requirements of a specified food safety/quality standard or scheme, that is, it confirms the compliance of the system to the agreed standard. To obtain certification, the implemented FSQS must be subjected to an independent audit by an accredited third party certification body and certified by them.[e] Following certification, surveillance audits will be conducted at a prescribed frequency and in accordance with a specified set of conditions to facilitate the maintenance of certification. The organization will be allowed to use the certification mark for the specific scheme[f] in its marketing efforts. Some organizations set out on a path of development and implementation of a FSQS with no plan to seek certification only to recognize later the tremendous benefits to be derived by maintaining a certified system.

5. Develop a FSQS communication plan

   Communication throughout the development, implementation, maintenance, and continual improvement of the FSQS is essential to ensuring that all stakeholders are kept informed of the status of the process. Sharing of information should not be done in an ad hoc manner, but should be a planned activity. Information that is developed for dissemination to stakeholders should be properly developed to convey the correct messages at an appropriate frequency. Different methods of communication may be used to achieve the objective of ensuring that all stakeholders are constantly aware of the status of the process and the role that they are required to play to advance it. Modes of communication that are utilized by the organization in the dissemination of other types of information should also be considered in the development of the communication strategy, for example, notice boards in high-traffic areas, intranet, internet, newsletters, departmental and general meetings, etc.

6. Explore various food safety and quality standards and schemes, pertinent legislation, and customers' requirements

The senior management team of an organization that is desirous of developing and implementing a FSQS should explore the available standards/schemes, relevant legislation, and specific customers' requirements in the process of determining the standard or scheme on which the FSQS should be based. In this regard, an organization may wish to consider standards or schemes that have been benchmarked by the GFSI and are therefore recognized. The countries in which the organization does business forms a key element of this review, as national legal requirements in the country in which products are packaged sometimes differ from requirements of countries to which the product is exported. It is critical that the FSQS meets the requirements of all countries in which the company serves customers. Some customers also have very specific requirements, which are enshrined in standards that they have developed and to which organizations are required to conform to be able to sustain a business relationship. Compliance with all relevant standards is therefore critical.

7. Select an appropriate food safety/quality standard/scheme

The selection of the standard or scheme on which the FSQS will be based is critical, as the established system must ensure the development and implementation of a system that will assure consistent conformance to the various quality and food safety requirements of the various markets in which the organization's products are sold and consumed. It is also imperative that the organization also develop a system to ensure that it is kept abreast of all amendments to standards/schemes to facilitate continuous conformance. Copies of all the various standards, schemes, legal, and customer requirements on which the FSQS is based should be kept on file.

8. Establish a food safety/quality team

Every employee in an organization plays a role in food safety and quality. To facilitate the development, implementation, maintenance, and continual improvement of the FSQS, a multidisciplinary team of employees should be appointed to coordinate the associated activities. This group of employees should represent key areas of the organization and should be equipped with the appropriate expertise to spearhead the development of a system that conforms to all the necessary requirements as set forth in the selected standard/scheme, legislation, and customers' requirements. The team members should include but not be limited to key functional areas, such as production, quality control, maintenance, distribution, human resource, and finance. Appointment of alternate members at the same time as the appointment of the core team members ensures that there is always a team of appropriate employees to facilitate the activities associated with the FSQS. The core team, as well as the alternates, should be exposed to appropriate training in the processes associated with the development, implementation, maintenance, and continual improvement of the FSQS. The appointment of the food safety/quality team should be formalized through documentation and notification of all staff of the members of the team and their roles.

9. Appoint a food safety/quality team leader

One of the food safety/quality team members, irrespective of his/her other duties within the organization, should be selected to perform the role of the team leader. The team leader would be required to manage the food safety team to ensure that the FSQS is developed and implemented in accordance with the plan and later maintained and continually improved over time. The team leader will also be responsible for ensuring that the food safety/quality team is adequately trained and provided with any other resources that are required to ensure that the team's objectives are met. Periodically, the food safety team member will also be required to update the leadership and management teams with reports on the functioning of the FSQS. Significant thought should therefore be injected into the selection of the team leader, as he/she will be required to play a critical role in the establishment of the FSQS. In many instances, the FSQS team leader (also sometimes called the FS/QS coordinator or manager) is the manager of the quality assurance department or a key member of that department. In all firms, depending on the organizational structure, this responsibility should be given to a senior manager with the authority to get things done or, at least to a person with direct access to such a manager that will ensure what is required gets done. In some firms, a very senior manager, vice president, or the equivalent is given direct responsibility for the FSQS. Whatever the choice of the firm, the FSQS will need to have the authority (either directly or indirectly), training.[g] The appointment of the food safety/quality team leader should be also formalized by way of documentation and information to all staff.

10. Provide the required resources to the food safety/quality team

The food safety/quality team must be provided with adequate resources to coordinate the development and implementation of the FSQS, as well as the maintenance and continual improvement activities. The most significant resource is training, which should be conducted prior to the commencement of the team's deliberations, as well as throughout the process on an as-needed basis. The FSQS team leader will be responsible for presenting the team's needs to the management team for approval. Adequate financial resources must also be made available in a timely manner to address all capital investment and other needs that may arise.

11. Conduct an official launch of the FSQS

The launch of the FSQS comprises an event or series of events in which employees and other stakeholders are advised of the commencement of the development and implementation of the company's FSMS and the implications for them. The launch may be supported by written communication in the form of articles on notice boards, the intranet, newsletter, etc. It is recommended that the launch meetings should be short and focused, that is, should not include other issues that are not associated with the FSQS.

It is recommended that the most senior executive or manager of the firm should facilitate the launch to convey the company's commitment to the food safety initiative. If this is not possible, another senior member of the management team could be selected to conduct the launch meeting(s), but the downside may well be a lack of certainty as to top management's full commitment to the program.

The launch presentation should be documented to ensure that the same message is conveyed to all employees and stakeholders and may include, but not be limited to the following topics:

a. strategic business objective to which the FSQS is tied,
b. strategic food safety objectives,
c. scope of the FSQS,
d. food safety policy,
e. commitment to certification or target for certification,
f. communication plan,
g. food safety/quality team and role of the food safety team,
h. food safety team leader and role of the food safety team leader,
i. employees' role in food safety, and
j. the FSQS project implementation plan.

12. Provide appropriate training for the food safety/quality team

Prior to the commencement of its activities, the food safety/quality team should be exposed to same basic food safety/quality awareness training that all other employees will receive to lay an appropriate foundation. On completion of the food safety/quality awareness training, the team should receive a more detailed training in food safety and quality, as well as the specifics of the food safety and quality standard or scheme, legal requirements, and customers' standards, which will form the basis of the FSQS. Training must be delivered by a competent individual and should include exit criteria. As the work of the team progresses, additional training for the FSQS team may be required. This should be brought to the attention of the senior management team to obtain the necessary approvals. Appropriate training records must be maintained for all training conducted for the food safety/quality team.

13. Create company-wide food safety/quality awareness

As mentioned previously, all employees play a role in the ensuring food safety and quality. To ensure that this is fully understood and embraced by all employees, food safety and quality awareness should be created company wide. This should involve the exposure of all employees, including senior executives, to presentations on food safety and quality and their role in ensuring that both are achieved. The awareness sessions could be relatively short sessions, which should include exit criteria (written or oral—but documented—tests for each attendee). The presentation should include but should not necessarily limited to the following:

a. definitions: food safety and quality,

b. food safety hazards,

c. food safety policy,

d. basis for FSQS (standard, scheme, legislation, customers' specifications, etc.),

e. GMP,

f. HACCP, and

g. FSQS implementation plan overview.

Key suppliers of contracted services (pest control, security, sanitation, etc.) should, as far as possible, also be exposed to the awareness training, as they are considered to be partners in the organization's food safety and quality initiative and need to be equally aware.

14. Conduct a gap analysis

A gap analysis is considered to be a critical feature of the development of the FSQS. The gap analysis is a method by which the differences between the organization's current food safety and quality arrangements and the requirements on which the FSQS will be based are identified. The gap analysis therefore identifies the opportunities for improvement with regard to food safety and quality. The gap analysis should be conducted by a competent individual and should be documented.

15. Evaluate and confirm the findings of the gap analysis

On completion of the gap analysis, the report should be carefully considered by the food safety/quality team. Clarification should be sought on any issues that are not particularly clear to the team. The gap analysis should be conducted relatively recently to ensure that the issues that need addressing are current.

16. Establish an approach to the development of a comprehensive FSQS

Based on the results of the gap analysis and the organization's target for the implementation of the FSQS, the food safety/quality team should propose an approach for the development of the leadership and management team. Issues, such as the aspects of the FSQS, areas of the operation, the products, the markets, etc., which should be addressed first should be carefully considered.

17. Develop a project plan for the development and implementation of the FSQS

Following careful consideration and approval of the proposed approach by the senior management team, a project plan should be developed by the food safety/quality team. This should include the various action steps, deliverables, responsibility, and targets, at the very least, for the development and the implementation of the FSQS. Activities may be scheduled to run concurrently to reduce the time for development and implementation. The process should include the documentation of policies, procedures, work instructions, as well as the preparation and deployment of all other required documents and records.

18. Approve the project plan
The plan should be approved by the senior management team as evidenced by signature and date. As the work of the food safety/quality team progresses, any changes to the project plan that need to be made should also be approved to ensure that appropriate focus is maintained.

19. Execute the project plan
The development and implementation activities should be executed in accordance with the project plan. Development includes the documentation of policies, procedures, and work instructions to facilitate conformance with the various requirements, while implementation involves the training of the relevant employees in the procedures and work instructions.

20. Develop an internal auditing capability
The company needs to ensure that it has the capability to monitor its conformance with agreed standards, regulations, practices, work instructions, and the overall FSQS being implemented. It will therefore have to ensure that its team leader and at least one other selected other team member are formally trained and certified in internal auditing. Equally important, is to ensure that it has the capability to deploy several multidepartmental teams or cross-functional teams to conduct periodic internal audits. As such, all members of the FSQS team, as well as other selected staff members should also be trained (perhaps internally) in auditing to the selected standard(s) and also according to the company's documented FSQS.

21. Communicate the status of development and implementation of the FSQS to key stakeholders
Any challenges encountered, which may impact the achievement of the specified targets, should be brought to the attention of senior management, as required. The necessary adjustments should be implemented. The food safety/quality team leader should provide the senior management team with periodic updates of progress and challenges at a prescribed frequency to ensure that management is always knowledgeable of the status of FSQS development and implementation.

22. Monitor conformance to and effectiveness of documented policies, procedures, and work instructions through periodic audits
Following the development and implementation of the policies, procedures, and work instructions, periodic internal audits may be conducted to determine not only conformance to the documentation but more importantly, but also the effectiveness of the system. The internal audits should be conducted by employees who have received appropriate training and who have been assessed as competent. A summary of results should be documented and submitted to the senior management team.

23. Conduct periodic scheduled management review meetings
Management review is a critical element of any comprehensive FSQS, as it represents an opportunity for the senior management team to consider progress and challenges in the different areas of the FSQS and the impact on the business. The management review meetings should be scheduled and attended by the senior management team (heads of functional areas). The food safety/quality team leader will be required to prepare internal food safety/quality reports for consideration by senior management.

24. Develop and implement corrective action plans
Based on the reports that are presented at management review meetings, the management team can determine the corrective actions, which should be employed to address the identified nonconformances and the associated support, which should be provided. The FSQS

team should then develop and implement corrective actions based on the results of the internal audits, any recent external audit, and management review meetings.

25. Arrange external audit of FSQS for certification

The external audit for certification can be arranged when the senior management team is satisfied that the FSQS is functioning effectively. The team may establish criteria for selection of an auditing company to conduct the audit and certify the organization.

26. Conduct certification audit

The certification audit should then be conducted in accordance with the predetermined arrangements. The organization's FSQS would be certified, providing that it meets the requirements for certification by the specific auditing company.

27. Develop corrective action plan based on the certification audit and implement corrective actions

A corrective action plan should be developed based on the results of the certification audit. The development of the plan should be a collaborative effort between the food safety team leader and all functional heads of areas in which nonconformances were identified. The corrective action plan should include the nonconformances, proposed corrective actions, responsibility, and target dates. The food safety team should conduct period evaluations to determine whether the corrective actions were implemented by the target dates so that they may be confirmed as complete.

28. Maintain certification

While an organization generally feels a sense of accomplishment having achieved certification, the challenge usually lies in maintenance of the FSQS. The organization is required to continue to conduct its operations in accordance with an effective documented system. To ensure that the organization maintains its certified system, surveillance audits are conducted by the auditor. Depending on the standard, scheme, legislation, or customer, surveillance audits may be announced or unannounced.

29. Continuous monitoring

Certification should not be considered to be the end point of the development and implementation of the FSQS. Periodic audits should continue at prescribed intervals to ensure that conformance to, and effectiveness of policies, procedures, and work instructions are maintained consistently. The results of these audits should be an agenda item at management review meetings and must be communicated to senior management for consideration.

30. Continually improve the FSQS

Based on the periodic internal audits that are conducted and other organizational activities, opportunities for improvement may be identified. The food safety team should develop plans for implementing the necessary modifications to the system to capitalize on the identified opportunities. The standards, schemes, legislation, and customers' requirements on which the FSQS is based may also be amended requiring a change within the organization's FSQS. These should also be accommodated in an attempt to continually improve the FSQS.

Quick wins

Here are a few ways to get some quick wins when implementing FSQS, which can then make the rest of the implementation process, which will take 8–18 months at best, completed efficiently and, more importantly, successfully.

- Develop and implement a GMP program prior to the implementation of a comprehensive FSQS based on a GFSI standard/scheme.

- Use any existing regulatory driver to full effect to "coerce" buy-in from both staff and senior management. *This tactic is one to be employed by the FSQS team/team leader to get a quick win as no one at any level is likely to risk the firms' business by resisting/obstructing changes required by law to continue operations.*[h]
- Document procedures and work instructions based on current practices, which can be modified later to facilitate conformance to the standard/scheme on which the FSQS is based. *Involving the production staff in this activity is a great idea that pays big dividends later on.*[i]
- For a company with several similar operations, develop documentation for one operation and use it as a template for the development of documentation for the others.
- Identify company personnel who can be trained as food safety/quality trainers and expose them to appropriate programs to facilitate internal training programs.
- Develop a team of internal auditors to conduct ongoing internal audits of the system as opposed to appointing a single internal auditor.

Experience had shown that focusing on these activities in the manner prescribed not only gains some quick, positive momentum, support, and buy-in from all concerned, but significantly shortens the implementation process and/or makes the process relative smooth.

Pitfalls

There are numerous likely pitfalls that may arise during the process of FSQS implementation in companies, wherever domiciled and these are more so evident in developing countries where resources tend to be more limited in most MSME's that are seeking to implement FSQ systems. These are caused by a wide variety of reasons, the main ones of which are summarized here.

1. Inadequate management commitment and involvement.
2. Commencing the project before the organization is truly ready or at an inappropriate time.
3. Lack of awareness of different standards/schemes, applicable legislation, or customers' requirements.
4. *Choosing a standard and an approach to FSQS implementation, which is not the most appropriate* given the type of organization.
5. Inadequate resources (financial and human).
6. Inadequate expert guidance.
7. PRPs not effectively implemented.
8. Infrastructural issues that require significant capital investment to resolve.
9. *Overemphasizing infrastructural issues versus systems implementation issues*, including documentation, record keeping, implementation of effective PRPs, implementing challenging but required changes to processing and practices, and developing and proving human resource competence.
10. Documentation developed by individuals who are not familiar with the processes or who do not understand the fundamentals or requirements of the FSQS being pursued.[j]
11. Complex documentation/creation of a large volume of paper work.
12. Time spent of creating perfect documentation.
13. Employees who need to use the documentation but are not trained in their use or do not understand them.
14. Documentation not made available to employees who need to use it.

15. Perception that the implementation of the FSQS means more work for the staff, rather than an improvement in practices.
16. Assignment of responsibility for the FSQS to an individual or group of individuals without adequate involvement of appropriate staff.[k]
17. Relaxation after certification and/or certification audits (inadequate maintenance).

These pitfalls often stymie successful implementation or lead to very extended implementation timelines. Whenever the implementation of a system is taking anywhere beyond 24 months, depending on the complexity of the system and the operations involved, it is likely that one, or more, or the aforementioned pitfalls are the reason. In this eventuality, implementing teams and the company's management need to take an impassioned look at the program, identify the issues, and deal with them.

Scientific information to support FSQS

The contents of the book, inclusive of the case studies presented, should have led to the conclusion that implementing compliant and sustainable FSQS or dealing effectively with market access issues resulting from food safety concerns is going to be highly dependent on the availability of competent scientific support. This necessity has been even further entrenched by the US Preventive Controls rule of the FDA's FSMA (Food and Drug Administration, 2015a) coming into the picture, which now requires scientifically valid, defensible proof of the efficacy of controls. This means the validation of supplier controls, allergen controls, sanitation controls (particularly for processes with no kill step), and process controls for all products [Food Safety Preventive Controls Alliance (FSCPA), 2016]. For traditional products from developing countries, this will mean the generation of scientific information to support their FSQS program and demonstrate the effectiveness of a range of controls involving ingredients, the specific product environment (e.g., gas-flushed packaging) combined methods processes (e.g., the Tanaka principle controls), and controls for suppliers. In this context, the information and approaches presented in this volume, combined with specific information of the microbiology, packaging, value chain management, and other considerations will be particularly relevant.

Endnotes

[a]This should be appropriate to the company, the level of sophistication of its production, warehousing and handling processes, its human resources, and, critically, its organizational culture (Yiannas, 2008).
[b]One of the most common reasons for failure is an inadequately resourced FSQS team.
[c]This is one of the more important activities because a gap audit that is not sufficiently thorough or which highlights important, but not critical or necessary issues as the main imperatives for action can lead to disillusionment and failure in implementation.
[d]A typical failing of technical personnel and technical advisors in developing country settings is not making sure that the top management or ownership of the businesses has a vested interest

in the outcome of the FSQS implementation process because it is not tied to achieving specific company goals. Examples of this would be accessing new markets, retaining existing markets, market expansion, marketing advantages in a competitive market, etc.

[e]Certification is *"a process by which accredited certification bodies, based on an audit, provide written assurance that food safety requirements and management systems and their implementation conformed to requirements"* (Global Food Safety Initiative, 2016b).

[f]See Chapter 2 for further information on the various GFSI schemes.

[g]Some schemes (e.g., Safe Quality Food) have very specific requirements for training of the food safety specialists (or SQF Practitioner, as SQF calls the individual). All schemes require the competence of persons with critical responsibilities within the FSQS, including the team leader, to be demonstrated by education, experience, on-the-job training, and/or other means. The FSMA in the United States also now specifically requires that the person signing off on all food safety and compliance documentation be trained and certified through a specific programme as a Preventive Controls Qualified Individual (PCQI). The Food Safety Preventive Control Alliance (FSCPA), formed with the support of the FDA, is responsible for certification [Food Safety Preventive Controls Alliance (FSCPA), 2016]. Also, the Food and Drug Act and SFCA in Canada and the Code of Federal Regulations (CFR) in the United States require individuals undertaking certain responsibilities to be trained and certified, for example, on a Better Process Control School (BPCS) for thermal processing of low-acid and acidified foods.

[h]An example of this is the implementation of the FSMA Preventive Controls, Produce, and FSVP rules, which have driven/are driving changes in months where none happened for years. While "coercion" is a strong term, its use is deliberate because, in essence, the implementors are *forcing* changes rather than getting voluntary buy-in.

[i]Involving the production team in documentation development can be done by a short meeting where the process/practices and documentation are discussed verbally and information/suggestions/observations recorded by a member of the technical team. This works well even in circumstances where there are literacy or language barrier issues, as it is a fully inclusive approach to making changes or documenting practices.

[j]This is a particular challenge that is often found where persons with a strong quality systems background (TQM, 5S, ISO 9001/14000) but very little scientific training or food safety training or understanding of food safety principles are providing technical guidance to firms. This will be further exacerbated by the ever-increasing requirements for competence under the SFCA, FSMA, and respective regulations that will continue to be implemented.

[k]It is quite common for the general staff of a firm/facility to be excluded from the development and implementation of the system, but are expected to embrace it, once complete. It is also quite common that the input of staff is not sought during the developmental stages or is ignored, despite the fact that they have day-to-day knowledge of the systems and practices that can determine the success or failure of controls beings developed and implemented.

References

Adalja, A.A., Vergis, E.N., 2010. *Actinomyces israelii* endocarditis misidentified as "Diptheroids". Anaerobe 16 (4), 472–473.

Aguilera, J.M., Arias, E.P., 1992. CYTED-D AHI: an Ibero American project on intermediate moisture foods and combined methods technology. Food Res. Int. 25 (2), 159–165.

Aguilera, J.M., Chirife, J., 1994. Combined methods for the preservation of foods in Latin America and the CYTED-D project. J. Food Eng. 22 (1), 433–444.

Ajandouz, E.H., Tchiakpe, L.S., Ore, F.D., Benajiba, A., Puigserver, A., 2001. Effects of pH on caramelization and Maillard reaction kinetics in fructose–lysine model systems. J. Food Sci. 66 (7), 926–931.

Al-Holy, M.A., Lin, M., Alhaj, O.A., Abu-Goush, M.H., 2015. Discrimination between *Bacillus* and *Alicyclobacillus* isolates in apple juice by Fourier Transform Infrared Spectroscopy and multivariate analysis. J. Food Sci. 80, M399–M404.

Andrews, S., Norton, I., Salunkhe, A.S., Goodluck, H., Aly, W.S.M., Mourad-Agha, H., Cornelis, P., 2013. Control of iron metabolism in bacteria. In: Banci, L. (Ed.), Metallomics and the Cell. Metal Ions in Life Sciences, vol. 12, Springer, New York (Chapter 7).

Appaiah, P., Sunil, L., Kumar, P.P., Krishna, A.G., 2015. Physico-chemical characteristics and stability aspects of coconut water and kernel at different stages of maturity. J. Food Sci. Tech. 52 (8), 5196–5203.

Awua, A.K., Doe, E.D., Agyare, R., 2011. Exploring the influence of sterilisation and storage on some physicochemical properties of coconut (*Cocos nucifera* L.) water. BMC Res. Notes 4 (1), 1.

Bélanger, P., Tanguay, F., Hamel, M., Phypers, M., 2015. An Overview of Foodborne Outbreaks in Canada Reported Through Outbreak Summaries: 2008–2014, CCDR, vol. 41–11, November 5, 2015: Foodborne Illness. Available from: http://www.phac-aspc.gc.ca/publicat/ccdr-rmtc/15vol41/dr-rm41-11/ar-01-eng.php

Bern, C., Hernandez, B., Lopez, M.B., Arrowood, M.J., de Mejia, M.A., de Merida, A.M., Hightower, A.W., Venczel, L., Herwaldt, B.L., Klein, R.E., 1999. Epidemiologic studies of *Cyclospora cayetanensis* in Guatemala. Emerg. Infect. Dis. 5 (6), 766.

Besser, R.E., Lett, S.M., Weber, J.T., Doyle, M.P., Barrett, T.J., Wells, J.G., Griffin, P.M., 1993. An outbreak of diarrhea and hemolytic uremic syndrome from *Escherichia coli* O157:H7 in fresh-pressed apple cider. JAMA 269 (17), 2217–2220.

Bevilacqua, A., Ciuffreda, E., Sinigaglia, M., Corbo, M.R., 2015. Spore inactivation and DPA release in *Alicyclobacillus acidoterrestris* under different stress conditions. Food Microbiol. 46, 299–306.

Bevilacqua, A., Corbo, M.R., Sinigaglia, M., 2008a. Inhibition of *Alicyclobacillus acidoterrestris* spores by natural compounds. Int. J. Food Sci. Technol. 43, 1271–1275.

Bevilacqua, A., Sinigaglia, M., Corbo, M.R., 2008b. *Alicyclobacillus acidoterrestris*: new methods for inhibiting spore germination. Int. J. Food Microbiol. 125, 103–110.

Bird, S., Traub, S.J., Grayzel, J., 2014. Organophosphate and carbamate poisoning. UpToDate, 14, 339. Available from: http://www.uptodate.com/contents/organophosphate-and-carbamate-poisoning

Food Safety and Quality Systems in Developing Countries. http://dx.doi.org/10.1016/B978-0-12-801226-0.00014-1

Blackburn, B.G., Craun, G.F., Yoder, J.S., Hill, V., Calderon, R.L., Chen, N., Lee, S.H., Levy, D.A., Beach, M.J., 2004. Surveillance for waterborne-disease outbreaks associated with drinking water—United States, 2001–2002. MMWR Surveill. Summ. 53 (8), 23–45.

Bottemiller, H., 2012. As *Salmonella* outbreak investigation continues, retailers recall mangoes from Mexico. More than 100 ill in *Salmonella* Braenderup outbreak. Available from: http://www.foodsafetynews.com/2012/08/as-salmonella-outbreak-investigation-continues-retailers-recall-mangoes-from-mexico/#.VwU0RFQrI_4

Breidt, F., Sandeep, K.P., Arrett, E.M., 2010. Use of linear models for thermal processing of acidified foods. Food Protect. Trends 30, 268–272.

Brown, M., Gordon, A., 1995. Use of accelerated shelf life testing in predicting the shelf life of fruit drinks. In: Institute of Food Technologists Annual Meeting, Anaheim, CA, June, 1995 (Abstract).

Buera, M., Chirife, J., Resnik, S.L., Lozano, R.D., 1987. Nonenzymatic browning in liquid model systems of high water activity: kinetics of color changes due to caramelization of various single sugars. J. Food Sci. 52 (4), 1059–1062.

Cai, R., Yuan, Y., Wang, Z., Wang, J., Yue, T., 2015. Discrimination of *Alicyclobacillus* strains by lipase and esterase fingerprints. Food Anal. Method 1, 1–6.

Calvin, L., Flores, L., Foster, W., 2003. Food safety in food security and food trade: case study—Guatemalan raspberries and *Cyclospora*. Focus 10, (Available from: http://lib.icimod.org/record/11312/files/4351.pdf).

Campos, C.F., Souza, P.E.A., Coelho, J.V., Glória, M.B.A., 1996. Chemical composition, enzyme activity and effect of enzyme inactivation on flavor quality of green coconut water. J. Food Process. Preserv. 20 (6), 487–500.

Canadian Food Inspection Agency (CFIA), 2016. Import requirements for fresh Guatemalan raspberries and blackberries. Available from: http://www.inspection.gc.ca/food/fresh-fruits-and-vegetables/imports-and-interprovincial-trade/guatemala-raspberries-and-blackberries/eng/1374597858728/1374597935222

Casey, P.G., Hill, C., Gahan, C.G., 2011. *E. coli* O104:H4: social media and the characterization of an emerging pathogen. Bioeng. Bugs 2 (4), 189–193.

Centers for Disease Control and Prevention (CDC), 1997. Update: outbreaks of cyclosporiasis—United States and Canada, 1997. Morb. Mortal. Wkly. Rep. 46 (23), 521.

Centers for Disease Control and Prevention (CDC), 2011. Multistate outbreak of human *Salmonella* Agona infections linked to whole, fresh imported papayas (Final Update). Available from: http://www.cdc.gov/salmonella/2011/papayas-8-29-2011.html

Centers for Disease Control and Prevention (CDC), 2012. Multistate outbreak of listeriosis linked to whole cantaloupes from Jensen Farms, Colorado (Final Update). Available from: http://www.cdc.gov/foodborneburden/index.html

Centers for Disease Control and Prevention (CDC), 2013. Outbreak of *Escherichia coli* O104:H4 infections associated with sprout consumption—Europe and North America, May–July 2011. Morb. Mortal. Wkly. Rep. 62 (50), 1029.

Centers for Disease Control and Prevention (CDC), 2016a. Estimated of foodborne illness in the United States. Available from: http://www.cdc.gov/foodborneburden/index.html

Centers for Disease Control and Prevention (CDC), 2016b. Foodborne germs and illnesses. Available from: http://www.cdc.gov/foodsafety/foodborne-germs.html

Cerny, G., Hennlich, W., Poralla, K., 1984. Fruchtsaftverderb durch bacillen: isolierung und charakterisierung des verderbserregers. Z. Lebensmitt. Unters. Forsch. 179, 224–227.

Chaine, A., Levy, C., Lacour, B., Riedel, C., Carlin, F., 2012. Decontamination of sugar syrup by pulsed light. J. Food Prot. 75 (5), 913–917.

Chang, S.S., Kang, D.H., 2004. *Alicyclobacillus* spp. in the fruit juice industry: history, characteristics, and current isolation/detection procedures. Crit. Rev. Microbiol. 30 (2), 55–74.

Ciuffreda, E., Bevilacqua, A., Sinigaglia, M., Corbo, M.R., 2015. *Alicyclobacillus* spp.: new insights on ecology and preserving food quality through new approaches. Microorganisms 3 (4), 625–640.

Clark, C.G., Price, L., Ahmed, R., Woodward, D.L., Melito, P.L., Rodgers, F.G., Jamieson, F., Ciebin, B., Li, A., Ellis, A., 2003. Characterization of waterborne outbreak–associated *Campylobacter jejuni*, Walkerton, Ontario. Emerg. Infect. Dis. 9 (10), 1232–1241.

Clarke-Harris, D., Reid, J., Fleischer, S., 1997. IPM Systems Development: Callaloo, *Amaranthus* sp. IPM CRSP: Fourth Annual Report 1996–1997, pp. 167–182.

Code of Federal Regulations, 2016a. Title 21 Food and Drugs Part 108. Emergency permit control. Available from: http://www.accessdata.fda.gov/scripts/cdrh/cfdocs/cfcfr/CFRSearch.cfm?CFRPart=108

Code of Federal Regulations, 2016b. Title 21 Food and Drugs Part 113. Thermally processed low-acid foods packaged in hermetically sealed containers. Available from: http://www.accessdata.fda.gov/scripts/cdrh/cfdocs/cfcfr/CFRSearch.cfm?CFRPart=113

Code of Federal Regulations, 2016c. Title 21 Food and Drugs Part 114. Acidified foods. Available from: http://www.accessdata.fda.gov/scripts/cdrh/cfdocs/cfcfr/CFRSearch.cfm?CFRPart=114

Codex Alimentarius Commission, 1985. Codex Alimentarius; General Principles of Food Hygiene. Recommended International Code of Practice (CAC/RCP1-1969, Rev. 2). Codex Alimentarius Commission, London, UK.

Codex Alimentarius, 1993. Low-Acid, O.S.I., Recommended International Code of Hygienic Practice for Low and Acidified Low Acid Canned Foods (CAC/RCP 23-1979, Rev. 2).

Cho, D.H., Lee, W.J., 1970. Microbiological studies of Korean native soy-sauce fermentation: a study on the microbora of fermented Korean Maeju loaves. J. Korean Soc. Appl. Biol. Chem. 13 (1), 35–42.

Chung, O., 2010. "A Sauce for All". Taiwan Review, Government Information Office, Republic of China, Taiwan, January 1, 2010.

Cooper, R., 2014. Travel giant First Choice pays out £1.7 million to 600 holidaymakers who fell ill at Turkish resort that had faeces in the pool. Available from: http://www.dailymail.co.uk/news/article-2561319/Travel-giant-First-Choice-pays-1-7million-holidaymakers-fell-ill-Turkish-resort-FAECES-pool.html

Cullimore, D.R., McCann, A.E., 1977. The Identification. Cultivation and Control of Iron Bacteria in Ground Water. Academic Press Inc, London, pp. 219–261.

da Costa, M.S., Rainey, F.A., 2010. Family II. Alicyclobacillaceae fam. nov. In: De Vos, P., Garrity, G., Jones, D., Krieg, N.R., Ludwig, W., Rainey, F.A., Schleifer, K.H., Whitman, W.B. (Eds.), Bergey's Manual of Systematic Bacteriology, vol. 3, second ed. Springer, New York, NY, USA, p. 229.

Damodaran, S., Parkin, K.L., Fennema, O.R. (Eds.), 2007. Fennema's Food Chemistry. CRC Press.

Davidson, A., 2004. Seafood of South-East Asia: A Comprehensive Guide With Recipes. Ten Speed Press, Penguin Random House, New York, p. 197.

Deinhard, G., Blanz, P., Poralla, K., Altan, E., 1987. *Bacillus acidoterrestris* sp. nov., a new thermotolerant acidophile isolated from different soils. Syst. Appl. Microbiol. 10, 47–53.

Den Hartog, C., 1978. In: Dixon, P.S., Irvine, L.M. (Eds.), Seaweeds of the British Isles, Volume 1, *Rhodophyta*. Part 1, Introduction, Nemaliales, Gigartinales. British Museum (Natural History), London, p. 252.

dos Anjos, M.M., Ruiz, S.P., Nakamura, C.V., de Abreu, F., Alves, B., 2013. Resistance of *Alicyclobacillus acidoterrestris* spores and biofilm to industrial sanitizers. J. Food Prot. 76 (8), 1408–1413.

do Nascimento Debien, I.C., de Moraes Santos Gomes, M.T., Ongaratto, R.S., Viotto, L.A., 2013. Ultrafiltration performance of PVDF, PES, and cellulose membranes for the treatment of coconut water (*Cocos nucifera* L.). Food Sci. Technol. 33 (4), 676–684.

Durand, P., 1996. Primary structure of the 16S rRNA gene of *Sulfobacillus thermosulfidooxidans* by direct sequencing of PCR amplified gene and its similarity with that of other moderately thermophilic chemolithotrophic bacteria. Syst. Appl. Microbiol. 19, 360–364.

ECLAC, 2011. The United States and Latin America and the Caribbean. Data on Economics and Trade. The Economic Commission for Latin America and the Caribbean (ECLAC). United Nations, Santiago, Chile, March, 2011.

Eckmanns, T., Oppert, M., Martin, M., Amorosa, R., Zuschneid, I., Frei, U., Rüden, H., Weist, K., 2008. An outbreak of hospital-acquired *Pseudomonas aeruginosa* infection caused by contaminated bottled water in intensive care units. Clin. Microbiol. Infect. 14 (5), 454–458.

Eiroa, M.N.U., Junquera, V.C.A., Schmidt, F.L., 1999. *Alicyclobacillus* in orange juice: occurrence and heat resistance of spores. J. Food Prot. 62, 883–886.

Euromonitor, 2011. Coconut water—a world of opportunity. Euromonitor Research. Euromonitor International Inc., Chicago, USA. Available from: http://blog.euromonitor.com/2011/06/coconut-water-a-world-of-opportunity.html

European Food Information Council (EUFIC), 2014. Viral foodborne illnesses. Available from: http://www.eufic.org/article/en/artid/Viral-foodborne-illnesses/

European Union Food Safety Authority (EUFSA), 2003. Commission Decision of 20 June 2003 on emergency measures regarding hot chilli and hot chilli products. Off. J. Eur. Union (2003/460/EC).

European Union Food Safety Authority (EUFSA), 2005. Commission Decision of 23 May 2005 on emergency measures regarding chilli products, curcuma and palm oil. Off. J. Eur. Union (2005/402/EC) L135/34.

Evans, M.R., Ribeiro, C.D., Salmon, R.L., 2003. Hazards of healthy living: bottled water and salad vegetables as risk factors for *Campylobacter* infection. Emerg. Infect. Dis. 9 (10), 1219.

FAO, 2000. New sports drink: coconut water. Agriculture 21 (11). Agriculture Department, Food and Agriculture Organization of the United Nations (FAO).

FAO, 2007. How to bottle coconut water. Spotlight, 2007. Available from: http://www.fao.org/ag/magazine/0701sp1.htm

FAOSTAT, 2014. Crop Production Statistics. The Statistics Division of the Food and Agriculture Organization of the United Nations, Rome. Available from: http://faostat3.fao.org/home/E

Fehrmann, H., Diamond, A.E., 1967. Peroxidase activity and *Phytophthora infestans* resistance in different organs of the potato plant. Phytopathology 57, 69–72.

Flynn, D., 2013. Last big *Cyclospora* outbreak was traced to Guatemalan raspberries. Food Safety News, July, 2013. Available from: http://www.foodsafetynews.com/2013/07/last-big-cyclospora-outbreak-was-traced-to-guatemalan-raspberries/#.Vzi-hoSDGko

Fondio, L., Grubben, G.J.H., 2004. Plant Resources of Tropical Africa 2. Vegetables. PROTA Foundation; Backhuys Publishes; CTA, Wageningen; Leiden; The Netherlands, pp. 217–221.

Food and Drug Administration, 2001. Hazard Analysis and Critical Control Point (HAACP); Procedures for the Safe and Sanitary Processing and Importing of Juice; Final Rule. Health and Human Services, Food and Drug Administration. Available from: https://www.gpo.gov/fdsys/pkg/FR-2001-01-19/pdf/01-1291.pdf

Food and Drug Administration, 2002. Guidance for Commercial Processors of Acidified and Low-Acid Canned Foods. Available from: http://www.fda.gov/Food/GuidanceRegulation/GuidanceDocumentsRegulatoryInformation/AcidifiedLACF/default.htm#Registration

Food and Drug Administration, 2012a. Instructions for Electronic Submission of Forms FDA 2541a and FDA 2541c (Process Filing Forms for Acidified and Low-Acid Foods) (AF/ LACF), Authorizing Access to the Electronic AF/LACF System, and Searching Process Filings. Food and Drug Administration Center for Food Safety and Applied Nutrition, September, 2012, US Department of Health and Human Services.

Food and Drug Administration, 2012b. 21 CFR 133 Cheese and Related Cheese Products. United States Government Publishing Office, Code of the Federal Register, p. 360. Available from: https://www.gpo.gov/fdsys/pkg/CFR-2012-title21-vol2/pdf/CFR-2012-title21-vol2-part133.pdf

Food and Drug Administration, 2014. Information on recalled Jensen Farms whole cantaloupe. Available from: http://www.fda.gov/Food/RecallsOutbreaksEmergencies/Outbreaks/ucm272372.htm

Food and Drug Administration, 2015a. FSMA final rule for preventive controls for human food. Current good manufacturing practice and hazard analysis and risk-based preventive controls for human food. Available from: http://www.fda.gov/downloads/Food/Guidance-Regulation/FSMA/UCM360735.pdf

Food and Drug Administration, 2015b. FSMA final rule on foreign supplier verification programs (FSVP) for importers of food for humans and animals. Available from: http://www.fda.gov/food/guidanceregulation/fsma/ucm361902.htm

Food and Drug Administration, 2015c. FSMA final rule on produce safety. Standards for the growing, harvesting, packing, and holding of produce for human consumption. Available from: http://www.fda.gov/Food/GuidanceRegulation/FSMA/ucm334114.htm

Food and Drug Administration, 2015d. FSMA final rule for preventive controls on human food: current good manufacturing practices and hazard analysis and risk-based preventive controls for human food. Available from: http://www.fda.gov/Food/GuidanceRegulation/FSMA/ucm334115.htm

Food and Drug Administration, 2015e. 21 CFR 133.169. Cheese and related cheese products. United States Government Publishing Office, Code of the Federal Register, pp. 360. Available from: https://www.gpo.gov/fdsys/pkg/CFR-2012-title21-vol2/pdf/CFR-2012-title21-vol2-part133.pdf

Food and Drug Administration, 2016a. Guidance for Commercial Processors of Acidified and Low-Acid Canned Foods. Available from: http://www.fda.gov/Food/GuidanceRegulation/GuidanceDocumentsRegulatoryInformation/AcidifiedLACF/default.htm#Registration

Food and Drug Administration, 2016b. FDA Industry Guidance Documents. Available from: http://www.fda.gov/downloads/Food/GuidanceRegulation

Food and Drug Administration, 2016c. FDA information for filing a cold fill and hold process, FDA Thermal Processing Guidance Regulation. Available from: http://www.fda.gov/downloads/Food/GuidanceRegulation/UCM252435.pdf

Food and Drug Administration, 2016d. Defect action levels in laboratory methods. Available from: http://www.fda.gov/food/foodscienceresearch/laboratorymethods/ucm083194.htm

Food Safety News, 2014. Kroger claims Primus has primary liability for deadly *Listeria* outbreak. Sues third party auditor used by Jensen Farms. Available from: http://www.foodsafetynews.com/2014/06/kroger-says-primus-has-primary-liability-for-deadly-outbreak/#.Vyw3SvkrK00

Food Safety Preventive Controls Alliance (FSCPA), 2016. Preventive Controls for Human Food, first ed. FSPCA, Bedford Park, Illinois, USA.

Food Standards Agency, 2016. New UK food poisoning figures published (last updated June 2014). Available from: https://www.food.gov.uk/news-updates/news/2014/6097/foodpoisoning

Foundation for Food Safety Certification, 2013. Strengths and Benefits. Note-11-4390-FSSC. Foundation for Food Safety Certification (FFSC), Gorinchem, The Netherlands. Available from: http://www.fssc22000.com/documents/support/downloads.xml?lang=en

Franciosa, G., Pourshaban, M., Gianfranceschi, M., Gattuso, A., Fenicia, L., Ferrini, A.M., Mannoni, V., De Luca, G., Aureli, P., 1999. *Clostridium botulinum* spores and toxin in mascarpone cheese and other milk products. J. Food Prot. 62 (8), 867–871.

Gault, G., Weill, F.X., Mariani-Kurkdjian, P., Jourdan-da Silva, N., King, L., Aldabe, B., Charron, M., Ong, N., Castor, C., Mace, M., Bingen, E., 2011. Outbreak of haemolytic uraemic syndrome and bloody diarrhoea due to Escherichia coli O104: H4, south-west France, June 2011. Euro Surveill. 16 (26), 19905.

Gavin, A., Weddig, L.M., 1995. Canned Foods: Principles of Thermal Process Control, Acidification and Container Closure Evaluation. Food Processors Institute, Washington, DC.

Glass, K., 2003. Tastee Process Cheese Spread Safety Evaluation. Food Research Institute, Wisconsin (for Dairy Industries Jamaica Limited).

Glass, K.A., Kaufman, K.M., Johnson, E.A., 1998. Survival of bacterial pathogens in pasteurized process cheese slices stored at 30°C. J. Food Prot. 61 (3), 290–294.

Glass, K.A., Johnson, E.A., 2004a. Factors that contribute to the botulinal safety of reduced-fat and fat-free process cheese products. J. Food Prot. 67 (8), 1687–1693.

Glass, K.A., Johnson, E.A., 2004b. Antibotulinal activity of process cheese ingredients. J. Food Prot. 67 (8), 1765–1769.

Global Food Forums, 2016. 2016 Food trends. Available from: http://www.globalfoodforums.com/food-news-bites/2016-food-trends/

Global Food Safety Initiative, 2016a. What is GFSI? Available from: http://www.mygfsi.com/about-us/about-gfsi/what-is-gfsi.html

Global Food Safety Initiative, 2016b. Third party certification and accreditation. Available from: http://www.mygfsi.com/schemes-certification/certification/third-party-certification-and-accreditation.html

Gordon, A., 2003. Evaluation of the Thermal Process Adequacy for Tastee Process Cheese Spread. Technological Solutions Limited, Kingston. Jamaica (for Dairy Industries Jamaica Limited).

Gordon, A. (Ed.), 2015a. Food Safety and Quality Systems in Developing Countries: Volume One: Export Challenges and Implementation Strategies. Academic Press, London, UK.

Gordon, A., 2015b. Dealing with trade challenges: science-based solutions to market-access interruption. In: Gordon, A. (Ed.), Food Safety and Quality Systems in Developing Countries: Volume One: Export Challenges and Implementation Strategies. Academic Press, London, UK, pp. 115–128.

Gordon, A., 2015c. Exporting Traditional fruits and vegetables to the United States: trade, food science, and sanitary and phytosanitary/technical barriers to trade considerations. In: Gordon, A. (Ed.), Food Safety and Quality Systems in Developing Countries: Volume One: Export Challenges and Implementation Strategies. Academic Press, London, UK, p. 4.

Gordon, A., Blake, G., Knight, O., 2011. Challenge Study on Tastee Cheese Produced by DIJL. Technological Solutions Limited, Kingston, Jamaica, (for Dairy Industries Jamaica Limited).

Gordon, A., Powell, M., Simpson, K., 2006. Challenge Study on DIJL Multivac Cheese. Technological Solutions Limited, Kingston, Jamaica (for Dairy Industries Jamaica Limited).

Gordon, A., Saltsman, J., Ware, G., Kerr, J., 2015a. Re-entering the US Market with Jamaican Ackees: a case study. In: Gordon, A. (Ed.), Food Safety and Quality Systems in Developing Countries: Volume One: Export Challenges and Implementation Strategies. Academic Press, London, UK, pp. 91–114.

Gordon, A., Williams, R., Gordon, T., Blake, S.G., Anderson, N., 2015b. Validation Studies on Dairy Industry Jamaica Limited's Canadian Canned Cheese. Technological Solutions Limited, Kingston, Jamaica.

Goreau, T.J., Trench, R.K., 2013. Innovative Methods of Marine Ecosystem Restoration. CRC Press, p. 193.

Goto, K., Mochida, K., Asahara, M., Suzuki, M., Kasai, H., Yokota, A., 2003. *Alicyclobacillus pomorum* sp. nov., a novel thermo-acidophilic, endospore-forming bacterium that does not possess ω-alicyclic fatty acids, and emended description of the genus *Alicyclobacillus*. Int. J. Syst. Evol. Microbiol. 53, 1537–1544.

Goto, K., Mochida, K., Kato, Y., Asahara, M., Fujita, R., An, S.Y., Kasai, H., Yokota, A., 2007. Proposal of six species of moderately thermophilic, acidophilic, endospore-forming bacteria: *Alicyclobacillus contaminans* sp. nov., *Alicyclobacillus fastidiosus* sp. nov., *Alicyclobacillus kakegawensis* sp. nov., *Alicyclobacillus macrosporangiidus* sp. nov., *Alicyclobacillus sacchari* sp. nov. and *Alicyclobacillus shizuokensis* sp. nov. Int. J. Syst. Evol. Microbiol. 57 (6), 1276–1285.

Guo, X., You, X.Y., Liu, L.J., Zhang, J.Y., Liu, S.J., Jiang, C.Y., 2009. *Alicyclobacillus aeris* sp. nov., a novel ferrous- and sulphur-oxidizing bacterium isolated from a copper mine. Int. J. Syst. Evol. Microbiol. 59, 2415–2420.

Gull, I., Saeed, M., Shaukat, H., Aslam, S.M., Samra, Z.Q., Athar, A.M., 2012. Inhibitory effect of *Allium sativum* and *Zingiber officinale* extracts on clinically important drug resistant pathogenic bacteria. Ann. Clin. Microbiol. Antimicrob. 11 (1), 1.

Hippchen, B., Roőll, A., Poralla, K., 1981. Occurrence in soil of thermo-acidophilic bacilli possessing ω-cyclohexane fatty acids and hopanoids. Arch. Microbiol. 129, 53–55.

Huang, X.C., Yuan, Y.H., Guo, C.F., Gekas, V., Yue, T.L., 2015. *Alicyclobacillus* in the fruit juice industry: spoilage, detection, and prevention/control. Food Rev. Int. 31 (2), 91–124.

Hünniger, T., Felbinger, C., Wessels, H., Mast, S., Hoffmann, A., Schefer, A., Märtlbauer, E., Paschke-Kratzin, A., Fischer, M., 2015. Food targeting: a real-time PCR assay targeting 16S rDNA for direct quantification of *Alicyclobacillus* spp. spores after aptamer-based enrichment. J. Agric. Food Chem. 63, 4291–4296.

Hach, 2015. Industry guide: bottled water. Available from: http://www.hach.com/bottledwater-guide.

Hahn, F., 2012. An on-line detector for efficiently sorting coconut water at four stages of maturity. Biosyst. Eng. 111 (1), 49–56.

Hallam, D., Liu, P., Lavers, G., Pilkauskas, P., Rapsomanikis, G., Claro, J., 2004. The market for non-traditional agricultural exports. Technical Paper, Raw Materials, Tropical and Horticultural Products Service Commodities and Trade Division, United Nations Food and Agricultural Organization, Rome, Italy, 2004.

Harrigan, W.F., McCance, M.E., 1976. Statistical methods for the selection and examination of microbial colonies. Laboratory Methods in Food and Dairy MicrobiologyAcademic Press, London, pp. 47–49.

Hayenga, A., 2011. Sudan red dye standards. AnalytiX, vol. 8, Article 4, SigmaAldrich, pp. 1–3. Available from: http://www.sigmaaldrich.com/technical-documents/articles/analytix/sudan-red-dye-standards.html#sthash.5TrYPBXq.dpuf

Herwaldt, B.L., Ackers, M.L., 1997. An outbreak in 1996 of cyclosporiasis associated with imported raspberries. N. Engl. J. Med. 336 (22), 1548–1556.

Heyliger, T.L., 2012. Thermal process calculations: overview of the General Method, Ball Formula and NumeriCAL. In: IFTPS South East Asia Technical Outreach Seminar, November 27, 2012, JBT Food Tech.

Horitsu, H., Wang, M.Y., Kawai, K., 1991. A modified process for soy sauce fermentation by immobilized yeasts. Agric. Biol. Chem. 55 (1), 269–271.

Hrudey, S.E., Payment, P., Huck, P.M., Gillham, R.W., Hrudey, E.J., 2003. A fatal waterborne disease epidemic in Walkerton, Ontario: comparison with other waterborne outbreaks in the developed world. Water Sci. Tech. 47 (3), 7–14.

IFTPS, 2016. Nomenclature for studies in thermal processing. Available from: http://www.iftps.org/pdf/nomenclature_6_04.pdf

Indar, L., 2012. Estimating the burden of foodborne diseases in the Caribbean. CAREC/PAHO/WHO. Available from: http://www.paho.org/hq/index.php?option=com_docman&task=doc_view&gid=15677&Itemid=2146

Institute of Food Technologists, 2015. Innova's top food, beverage trends for 2016. Available from: http://www.ift.org/Food-Technology/Daily-News/2015/November/17/Innovas-top-10-food-beverage-trends-for-2016.aspx

International Bottled Water Association, 2005. Bottled water code of practice. Available from: http://www.bottledwater.org/public/pdf.IBWA05ModelCode_Mar2.pdf.

International Food Information Council (IFIC) Foundation, 2013. 2013 Food and health survey. Available from: www.foodinsights.com

ITC TradeMap, 2015. List of exporters for the selected product in 2014. Product: 2201 mineral and aerated waters. Available from: http://www.trademap.org/Country_SelProduct.aspx?nvpm=1|||||2201|||4|1|1|2|1||2|1|1

ITC, 2016. Market Analysis and Research From Trade Map. International Trade Centre (ITC). Geneva, Switzerland.

Jackson, J., 2002. Coconut Water Quality Evaluation. Final Report submitted to the Food and Agriculture Organization of the United Nations (FAO), Rome.

Jackson, J.C., Gordon, A., Wizzard, G., McCook, K., Rolle, R., 2004. Changes in chemical composition of coconut (Cocos nucifera) water during maturation of the fruit. J. Sci. Food Agric. 84 (9), 1049–1052.

Jensen, N., Whitfield, F.B., 2003. Role of Alicyclobacillus acidoterrestris in the development of a disinfectant taint in shelf-stable fruit juice. Lett. Appl. Microbiol. 36 (1), 9–14.

Jiang, C.Y., Liu, Y., Liu, Y.Y., You, X.Y., Guo, X., Liu, S.J., 2008. Alicyclobacillus ferrooxydans sp. nov., a ferrous-oxidizing bacterium from solfataric soil. Int. J. Syst. Evol. Microbiol. 58, 2898–2903.

Johnson, M.E., Kapoor, R., McMahon, D.J., McCoy, D.R., Narasimmon, R.G., 2009. Reduction of sodium and fat levels in natural and processed cheeses: scientific and technological aspects. Compr. Rev. Food Sci. Food Saf. 8 (3), 252–268.

Kataoka, S., 2005. Functional effects of Japanese style fermented soy sauce (shoyu) and its components. J. Biosci. Bioeng. 100 (3), 227–234.

Kawase, K.Y., Luchese, R.H., Coelho, G.L., 2013. Micronized benzoic acid decreases the concentration necessary to preserve acidic beverages against Alicyclobacillus. J. Appl. Microbiol. 115 (2), 466–474.

Knight, O., Gordon, C. L. A., 2004. Alicyclobacillus in tropical Caribbean beverages: incidents and intervention strategies for control of this threat to the industry. In: 2004 IFT Annual Meeting, 12–16 July, 2004, Las Vegas, NV.

Kitchin, C., 2015. British holidaymakers take legal action against Thomson Holidays after being diagnosed with irritable bowel syndrome following dream trip to Dominican Republic. MailOnline. Available from: http://www.dailymail.co.uk/travel/travel_news/article-3215894/Dirk-Ord-Stephen-Robson-sue-Thomson-illness-Dominican-Republic-hotel.html

Komitopoulou, E., Boziaris, I.S., Davies, E.A., Delves-Broughton, J., Adams, M.R., 1999. Alicyclobacillus acidoterrestris in fruit juices and its control by nisin. Int. J. Food Sci. Technol. 34 (1), 81–85.

Kovats, S.K., Doyle, M.P., Tanaka, N., 1984. Evaluation of the microbiological safety of tofu. J. Food Prot. 47 (8), 618–622.

Kowalski, S., Lukasiewicz, M., Duda-Chodak, A., Zięć, G., 2013. 5-Hydroxymethyl-2-furfural (HMF)–heat-induced formation, occurrence in food and biotransformation–a review. Polish J. Food Nutr. Sci. 63 (4), 207–225.

Labuza, T.P., 1982. Shelf-Life Dating of Foods. Food and Nutrition Press, Wesport, CT.

Labuza, T.P., 1984. Application of chemical kinetics to deterioration of foods. J. Chem. Educ. 61, 348–358.

Labuza, T.P., Schmidl, M.K., 1985. Accelerated shelf-life testing. Food Technol. 39, 57–64, p. 134.

Lam, S., Samraj, J., Rahman, S., Hilton, E., 1993. Primary actinomycotic endocarditis: case report and review. Clin. Infect. Dis. 16 (4), 481–485.

Lazarkova, Z., Buňka, F., Buňková, L., Holáň, F., Kráčmar, S., Hrabě, J., 2011. The effect of different heat sterilization regimes on the quality of canned processed cheese. J. Food Process Eng. 34 (6), 1860–1878.

Leclerc, H., Schwartzbrod, L., Dei-Cas, E., 2002. Microbial agents associated with waterborne diseases. Crit. Rev. Microbiol. 28 (4), 371–409.

Lee, W.J., Cho, D.H., 1971. Microbiological studies of Korean native soy-sauce fermentation— A study on the microflora changes during Korean native soy-sauce fermentation. J. Korean Soc. Appl. Biol. Chem. 14 (2), 137–148.

Lee, H.S., Nagy, S., 1990. Formation of 4-vinyl guaiacol in adversely stored orange juice as measured by an improved HPLC method. J. Food Sci. 55 (1), 162–163.

Leistner, L., 1994. Further developments in the utilization of hurdle technology for food preservation. J. Food Eng. 22 (1), 421–432.

Leistner, L., 2000. Basic aspects of food preservation by hurdle technology. Int. J. Food Microbiol. 55 (1), 181–186.

Leistner, L., Rödel, W., 1976. Inhibition of Micro-organisms in Food by Water Activity. Academic Press, London, UK.

Leyer, G.J., Wang, L.L., Johnson, E.A., 1995. Acid adaptation of *Escherichia coli* O157:H7 increases survival in acidic foods. Appl. Environ. Microbiol. 61 (10), 3752–3755.

Li, J., Huang, R., Xia, K., Liu, L., 2014. Double antibodies sandwich enzyme-linked immunosorbent assay for the detection of *Alicyclobacillus acidoterrestris* in apple juice concentrate. Food Control 40, 172–176.

Liu, T., Zhang, X., Zhu, W., Liu, W., Zhang, D., Wang, J., 2014. A G-quadruplex DNAzyme-based colorimetric method for facile detection of *Alicyclobacillus acidoterrestris*. Analyst 139 (17), 4315–4321.

Lund, B.M., 2015. Microbiological food safety for vulnerable people. Int. J. Environ. Res. Public Health 12 (8), 10117–10132.

Marler, B., 2011. 97 *Salmonella* illnesses linked to papayas from Mexico. Available from: http://www.foodpoisonjournal.com/foodborne-illness-outbreaks/97-salmonella-illnesses-linked-to-papayas-from-mexico/#.VwUnJFQrI_4

Marler, B., 2012. What have we learned from past *Salmonella* mango outbreaks? Not much. Available from: http://www.foodpoisonjournal.com/foodborne-illness-outbreaks/what-have-we-learned-from-past-salmonella-mango-outbreaks-not-much/#.VwUzh1QrI_4

Marler, B., 2014. Publisher's platform: three years since people died from cantaloupe Primus' "scorched earth" litigation strategy will change audit industry, retailer responsibility next. Food Safety News, July 22, 2014. Available from: http://www.foodsafetynews.com/2014/07/publishers-platform-three-years-since-the-primus-jensen-farms-audit/#.Vyw6S_krK00

Mahayothee, B., Koomyart, I., Khuwijitjaru, P., Siriwongwilaichat, P., Nagle, M., Müller, J., 2015. Phenolic compounds, antioxidant activity, and medium chain fatty acids profiles of coconut water and meat at different maturity stages. Int. J. Food Prop. 19 (9), 2014–2051.

Maheshwari, R.K., Singh, A.K., Gaddipati, J., Srimal, R.C., 2006. Multiple biological activities of curcumin: a short review. Life Sci. 78 (18), 2081–2087.

Mahnot, N.K., Kalita, D., Mahanta, C.L., Chaudhuri, M.K., 2014. Effect of additives on the quality of tender coconut water processed by nonthermal two stage microfiltration technique. LWT Food Sci. Technol. 59 (2), 1191–1195.

Martins, R.C., Lopes, V.V., Vicente, A.A., Teixeira, J.A., 2008. Computational shelf-life dating: complex systems approaches to food quality and safety. Food Bioprocess Technol. 1 (3), 207–222.

Matsubara, H., Goto, K., Matsumura, T., Mochida, K., Iwaki, M., Niwa, M., Yamasoto, K., 2002. *Alicyclobacillus acidiphilus* sp. nov., a novel thermo-acidophilic, ω-alicyclic fatty acid-containing bacterium isolated from acidic beverages. Int. J. Syst. Evol. Microbiol. 52, 1681–1685.

McGlynn, W., 2015. Guidelines for the Use of Chlorine Bleach as a Sanitizer in Food Processing Operations. Oklahoma State University Food Technology Fact Sheet, FAPC-116.

McKnight, I.C., Eiroa, M.N.U., Sant'Ana, A.S., Massaguer, P.R., 2010. *Alicyclobacillus acidoterrestris* in pasteurized exotic Brazilian fruit juices: isolation, genotypic characterization and heat resistance. Food Microbiol. 27 (8), 1016–1022.

Mendonca, A., 2006. Microbial Challenge Studies to Evaluate the Viability of Four Human Enteric Pathogens in Walkerswood Jamaican Jerk Seasoning at 25°C. Iowa State University, October 8, 2006. Submitted to Walkerswood Caribbean Foods Limited, St. Ann, Jamaica.

Meyer, R.S., DuVal, A.E., Jensen, H.R., 2012. Patterns and processes in crop domestication: an historical review and quantitative analysis of 203 global food crops. New Phytologist 196 (1), 29–48.

Mitchell, S.A., 2011. The Jamaican root tonics: a botanical reference. Focus Altern. Complement. Ther. 16 (4), 271–280.

Minnesota Department of Health, 2015. Iron bacteria in well water. Available from: http://www.health.state.mn.us/divs/eh/wells/waterquality/ironbacteria.html

Monro, J.A., Harding, W.R., Russell, C.E., 1985. Dietary fibre of coconuts from a pacific atoll: soluble and insoluble components in relation to maturity. J. Sci. Food Agric. 36 (10), 1013–1018.

Nair, R.V., 2004. Controversial Drug Plants. Universities Press, Hyderguda, Hyderabad, India, pp. 224–225.

Neto, U.F., Franco, L., Tabacow, K.M.B.D., Machado, N.L., 1993. Negative findings for use of coconut water as an oral rehydration solution in childhood diarrhea. J. Am. Coll. Nutr. 12 (2), 190–193.

Nielsen, S.S. (Ed.), 1998. Food Analysis, second ed. Aspen Publishers, Gaithersburg, MD, (Chapter 2).

New South Wales Government, 2016. Food poisoning. Department of Primary Industries Food Authority. Available from: http://www.foodauthority.nsw.gov.au/fp/food-poisoning

Nutrition Data, 2016. Foods highest in Vitamin C. Available from: http://nutritiondata.self.com/foods-000101000000000000000.html

NSF, 2015. NSF International Certification Requirements NSF International Certification Requirements for Bottled Water and Packaged Beverages August 3, 2015. NSF International, Ann Arbor, MI, USA.

Offley, S., 2016. Food tracker. Lethality calculations. Available from: http://support.fluke.com/datapaq/Download/Asset/9290118_ENG_A_W.PDF

Okereke, A., Montville, T.J., 1991a. Bacteriocin inhibition of *Clostridium botulinum* spores by lactic acid bacteria. J. Food Prot. 54 (5), 349–356.

Okereke, A., Montville, T.J., 1991b. Bacteriocin-mediated inhibition of *Clostridium botulinum* spores by lactic acid bacteria at refrigeration and abuse temperatures. Appl. Environ. Microbiol. 57 (12), 3423–3428.

Orr, R.V., Beuchat, L.R., 2002. Efficiency of disinfectants in killing spores of *Alicyclobacillus acidoterrestris* and performance of media for supporting colony development by survivors. J. Food Prot. 63, 1117–1122.

Ortega, Y.R., Sterling, C.R., Gilman, R.H., Cama, V.A., Diaz, F., 1993. *Cyclospora* species—a new protozoan pathogen of humans. N. Engl. J. Med. 328 (18), 1308–1312.

Ortega, Y.R., Sanchez, R., 2010. Update on *Cyclospora cayetanensis*, a food-borne and water-borne parasite. Clin. Microbiol. Rev. 23 (1), 218–234.

Oshima, M., Ariga, T., 1975. ω-Cyclohexyl fatty acids in acidophilic thermophilic bacteria. J. Biol. Chem. 250, 6963–6968.

O'Connor, D.R., Ontario Ministry of the Attorney General, 2002. Report of the Walkerton Inquiry: The Events of May 2000 and Related Issues—A Summary. Ontario Ministry of the Attorney General, Ontario, Canada.

Oteiza, J.M., Ares, G., Sant'Ana, A.S., Soto, S., Giannuzzi, L., 2011. Use of a multivariate approach to assess the incidence of *Alicyclobacillus* spp. in concentrate fruit juices marketed in Argentina: results of a 14-year survey. Int. J. Food Microbiol. 151 (2), 229–234.

Oteiza, J.M., Soto, S., Alvarenga, V.O., Sant'Ana, A.S., Giannuzzi, L., 2014. Flavorings as new sources of contamination by deteriogenic *Alicyclobacillus* of fruit juices and beverages. Int. J. Food Microbiol. 172, 119–124.

Oteiza, J.M., Soto, S., Alvarenga, V.O., Sant'Ana, A.S., Gianuzzi, L., 2015. Fate of *Alicyclobacillus* spp. in enrichment broth and in juice concentrates. Int. J. Food Microbiol. 210, 73–78.

Paish, M., 2011. Food safety week focus on new statistics releases and claims. Australia Food Safety News. Available from: http://ausfoodnews.com.au/2011/11/07/food-safety-week-focuses-on-new-statistics-releases-and-claims.html

Pei, J., Yue, T., Yuan, Y., 2014. Control of *Alicyclobacillus acidoterrestris* in fruit juices by a newly discovered bacteriocin. World J. Microbiol. Biotechnol. 30 (3), 855–863.

Palop, A., Lvarez, I., Raso, J., Condon, S., 2000. Heat resistance of *Alicyclobacillus acidocaldarius* in water, various buffers, and orange juice. J. Food Prot. 63, 1377–1380.

Pettipher, G.L., Osmundsen, M.E., Murphy, J.M., 1997. Methods for the detection, enumeration and identification of *Alicyclobacillus acidoterrestris* and investigation of growth and production of taint in fruit juice-containing drinks. Lett. Appl. Microbiol. 24, 185–189.

Pires, S.M., Vieira, A.R., Perez, E., Wong, D.L.F., Hald, T., 2012. Attributing human foodborne illness to food sources and water in Latin America and the Caribbean using data from outbreak investigations. Int. J. Food Microbiol. 152 (3), 129–138.

Piskernik, S., Klančnik, A., Demšar, L., Možina, S.S., Jeršek, B., 2016. Control of *Alicyclobacillus* spp. vegetative cells and spores in apple juice with rosemary extracts. Food Control 60, 205–214.

Ponting, J.D., Joslyn, M.A., 1948. Ascorbic acid oxidation and browning in apple tissue extracts. Arch. Biochem. 19, 47–63.

Pontius, A.J., Rushing, J.E., Foegeding, P.M., 1998. Heat resistance of *Alicyclobacillus acidoterrestris* spores as affected by various pH values and organic acids. J. Food Prot. 61 (1), 41–46.

Prades, A., Dornier, M., Diop, N., Pain, J.P., 2012. Coconut water uses, composition and properties: a review. Fruits 67 (2), 87–107.

Pradeepkumar, T., Sumajyothibhaskar, B., Satheesan, K.N., 2008. Management of Horticultural Crops. Horticulture Science Series, vol. 11, 2 parts. New India Publishing, New Delhi, India, pp. 539–587.

Pue, A.G., Rivu, W., Sundarrao, K., Kaluwin, C., Singh, K., 1992. Preliminary studies on changes in coconut water during maturation of the fruit. Sci. New Guinea 18, 81–84.

Qin, J., Cui, Y., Zhao, X., Rohde, H., Liang, T., Wolters, M., Li, D., Campos, C.B., Christner, M., Song, Y., Yang, R., 2011. Identification of the Shiga toxin-producing Escherichia coli O104:H4 strain responsible for a food poisoning outbreak in Germany by PCR. J. Clin. Microbiol. 49 (9), 3439–3440.

Rao, V.S., Naveen Kumar, R., Polasa, K., 2012. Foodborne diseases in India—a review. Br. Food J. 114 (5), 661–680.

Rao, T.S., Sairam, T.N., Viswanathan, B., Nair, K.V.K., 2000. Carbon steel corrosion by iron oxidising and sulphate reducing bacteria in a freshwater cooling system. Corros. Sci. 42 (8), 1417–1431.

Rapid Alert System for Food and Feed, 2016. Available from: http://ec.europa.eu/food/food/rapidalert/indexen.htm

Riley, L.W., Remis, R.S., Helgerson, S.D., McGee, H.B., Wells, J.G., Davis, B.R., Hebert, R.J., Olcott, E.S., Johnson, L.M., Hargrett, N.T., Blake, P.A., 1983. Hemorrhagic colitis associated with a rare Escherichia coli serotype. N. Engl. J. Med. 308 (12), 681–685.

Roberts, T.A., Bryan, F.L, Christian, J.H.B., Kilsby, D., Olson, Jr., J.C., Silliker, J.H., 1986. Microorganisms in Foods 2—Sampling for Microbiological Analysis, Principles and Specific Applications, second ed University of Toronto Press, Toronto, Canada.

Röling, W.F., Apriyantono, A., Van Verseveld, H.W., 1996. Comparison between traditional and industrial soy sauce (kecap) fermentation in Indonesia. J. Ferment. Bioeng. 81 (3), 275–278.

Rolle, R.S., 2007. Good Practice for the Small-Scale Production of Bottled Coconut Water, vol. 1. FAO Agricultural and Food Engineering Training and Resource Materials, Food and Agriculture Organization of the United Nations, Rome.

Romero-Frias, X., 2003. The Maldive Islanders: A Study of the Popular Culture of an Ancient Ocean Kingdom. Nova Ethnographia Indica, Barcelona, Spain.

Rothschild, M., 2011. More papayas found contaminated with Salmonella. Available from: http://www.foodsafetynews.com/2011/07/more-tainted-papaya-found-in-outbreak-investigation/#.VwUoLlQrI_4

Rouseff, R.L., Dettweiler, G.R., Swaine, R.M., Naim, M., Zehavi, U., 1992. Solid-phase extraction and HPLC determination of 4-vinyl guaiacol and its precursor, ferulic acid, in orange juice. J. Chromatogr. Sci. 30 (10), 383–387.

Rubino, S., Cappuccinelli, P., Kelvin, D.J., 2011. Escherichia coli (STEC) serotype O104 outbreak causing haemolytic syndrome (HUS) in Germany and France. J. Infect. Dev. Ctries. 5 (6), 437–440.

Saltsman, J., Gordon, A., 2015. The food safety modernization act and its impact on the Caribbean's approach to export market access. In: Gordon, A. (Ed.), Food Safety and Quality Systems in Developing Countries: Export Challenges and Implementation Strategies, vol. I, Academic Press, London, UK, pp. 129–145.

Sanders, L., 2015. 5 Tasty trends for 2016. International Food Information Council (IFIC) Foundation. Available from: http://www.foodinsight.org/2016-food-fads-trend

Saravanamuthu, R., 2010. Industrial Exploitation of Microorganisms. I.K. International Publication House, New Delhi, (p. 242).

Sloan, E., 2014. Top ten functional food trends. Food Technol. 68 (4).

Shemesh, M., Pasvolsky, R., Zakin, V., 2014. External pH is a cue for the behavioral switch that determines surface motility and biofilm formation of Alicyclobacillus acidoterrestris. J. Food Prot. 77, 1418–1423.

Shivsharan, U.S., Bhitre, M.J., 2013. Antimicrobial activity of lactic acid bacteria isolates. Int. Res. J. Pharm. 4 (8), 197–201.

Shurtleff, W., Aoyagi, A., 2009a. History of Soybeans and Soyfoods in the Caribbean/West Indies (1767–2008), Electronic Resource.

Shurtleff, W., Aoyagi, A., 2009b. History of Soybeans and Soyfoods in South America (1884–2009): Extensively Annotated Bibliography and Sourcebook. Soyinfo Center.

Shurtleff, W., Aoyagi, A., 2010. History of Soybeans and Soyfoods in Southeast Asia (13th Century to 2010): Extensively Annotated Bibliography and Sourcebook. Soyinfo Center.

Shurtleff, W., Aoyagi, A., 2012. History of Soy Sauce (160 CE to 2012). Soyinfo Center, Lafayette, CA.

Silva, F.V., Gibbs, P., 2001. *Alicyclobacillus acidoterrestris* spores in fruit products and design of pasteurization processes. Trends Food Sci. Technol. 12 (2), 68–74.

Silva, F.V.M., Gibbs, P., 2004. Target selection in designing pasteurization processes for shelf-stable high-acid fruit products. Crit. Rev. Food Sci. Nutr. 44, 353–360.

Silva, F.M., Gibbs, P., Vieira, M.C., Silva, C.L., 1999. Thermal inactivation of *Alicyclobacillus acidoterrestris* spores under different temperature, soluble solids and pH conditions for the design of fruit processes. Int. J. Food Microbiol. 51 (2), 95–103.

Silva, F.V., Tan, E.K., Farid, M., 2012. Bacterial spore inactivation at 45–65°C using high pressure processing: study of *Alicyclobacillus acidoterrestris* in orange juice. Food Microbiol. 32 (1), 206–211.

Singh, R., 2000. Scientific principles of shelf-life evaluation. In: Man, C.M.C., Jones, A.A. (Eds.), Shelf-Life Evaluation of Foods. Aspen, Gaithersburg, MD, pp. 3–23.

Singh, T.K., Cadwallader, K.R., 2004. Ways of measuring shelf-life and spoilage. Understanding and Measuring the Shelf-Life of FoodWoodhead Publishing, 165–183.

Speck, M.L., 1992. Compendium of methods for the microbiological examination of foods (No. QR115. C66 1992). APHA Technical Committee on Microbiological Methods for Foods.

Smith, A.H., Nichols, K., McLachlan, J., 2012. Cultivation of seamoss (*Graciliara*) in St. Lucia, West Indies. In: Bird, C.J., Ragan, M. (Eds.), Eleventh International Seaweed Symposium: Proceedings of the Eleventh International Seaweed Symposium. 19–25 June, 1983, Qingdao, Peoples Republic of China, pp. 249–251.

Somers, E.B., Taylor, S.L., 1987. Antibotulinal effectiveness of nisin in pasteurized process cheese spreads. J. Food Prot. 50 (10), 842–848.

Splittstoesser, D.F., Churey, J.J., Lee, Y., 1994. Growth characteristics of aciduric sporeforming *Bacilli* isolated from fruit juices. J. Food Prot. 57, 1080–1083.

Stumbo, C.R., 1973. Thermobacteriology in Food Processing. Academic Press, Inc, San Diego, CA.

Subramanjan, T.L., Vasudevan, P., 1977. A note on the physical and chemical composition of tender coconut water. Madras Agric. J. 64 (9), 616–617.

Sudershan, R.V., Naveen Kumar, R., Kashinath, L., Bhaskar, V., Polasa, K., 2014. Foodborne infections and intoxications in Hyderabad India. Epidemiol. Res. Int. 2014, (Article ID 942961, 5 pages).

Suezawa, Y., Kimura, I., Inoue, M., Gohda, N., Suzuki, M., 2006. Identification and typing of miso and soy sauce fermentation yeasts, *Candida etchellsii* and *C. versatilis*, based on sequence analyses of the D1D2 domain of the 26S ribosomal RNA gene, and the region of internal transcribed spacer 1, 5.8S ribosomal RNA gene and internal transcribed spacer 2. Biosci. Biotechnol. Biochem. 70 (2), 348–354.

Takeda, I., 2011. Recycling of Phosphorus Resources in Agricultural Areas Using Woody Biomass and Biogenic Iron Oxides. InTech. Available from: http://www.intechopen.com/books/biomass-detection-production-and-usage/recycling-of-phosphorus-resources-in-agricultural-areas-using-woody-biomass-and-biogenic-iron-oxides#exportas

Tamime, A.Y., 2011. Processed cheese and analogues: an overview. Processed Cheese and Analogues Blackwell Publishing Ltd, Oxford, UK.

Tan, B.K., Vanitha, J., 2004. Immunomodulatory and antimicrobial effects of some traditional Chinese medicinal herbs: a review. Curr. Med. Chem. 11 (11), 1423–1430.

Tianli, Y., Jiangbo, Z., Yahong, Y., 2014. Spoilage by *Alicyclobacillus* bacteria in juice and beverage products: chemical, physical, and combined control methods. Compr. Rev. Food Sci. Food Saf. 13 (5), 771–797.

Tanaka, N., 1982. Toxin production by *Clostridium botulinum* in media at pH lower than 4.6. J. Food Prot. 45 (3), 234–237.

Tanaka, Y., Nguyen, V.K., 2007. Edible Wild Plants of Vietnam. Orchid Press, Hong Kong, p. 24.

Tanaka, N., Goepfert, J.M., Traisman, E., Hoffbeck, W.M., 1979. A challenge of pasteurized process cheese spread with *Clostridium botulinum* spores. J. Food Prot. 42 (10), 787–789.

Tanaka, N., Kovats, S.K., Guggisberg, J.A., Meske, L.M., Doyle, M.P., 1985. Evaluation of the microbiological safety of tempeh made from unacidified soybeans. J. Food Prot. 48 (5), 438–441.

Tanaka, N., Traisman, E., Plantinga, P., Finn, L., Flom, W., Meske, L., Guggisberg, J., 1986. Evaluation of factors involved in antibotulinal properties of pasteurized process cheese spreads. J. Food Prot. 49 (7), 526–531.

Taylor, J.L., Tuttle, J., Pramukul, T., Brien, K.O., Barrett, T.J., Jolbitado, B., Lim, Y.L., Vugia, D., Morris, J.G., Tauxe, R.V., Dwyer, D.M., 1993. An outbreak of cholera in Maryland associated with imported commercial frozen fresh coconut milk. J. Infect. Dis. 167 (6), 1330–1335.

Technological Solutions Limited, 2007. Unpublished results.

Ter Steeg, P.F., Cuppers, H.G., Hellemons, J.C., Rijke, G., 1995. Growth of proteolytic *Clostridium botulinum* in process cheese products: I. Data acquisition for modeling the influence of pH, sodium chloride, emulsifying salts, fat dry basis, and temperature. J. Food Prot. 58 (10), 1091–1099.

Ter Steeg, P.F., Cuppers, H.G., 1995. Growth of proteolytic *Clostridium botulinum* in process cheese products: II. Predictive modeling. J. Food Prot. 58 (10), 1100–1108.

Todd, E.C., 1989. Costs of acute bacterial foodborne disease in Canada and the United States. Int. J. Food Microbiol. 9 (4), 313–326.

Tourova, T.P., Poltoraus, A.B., Lebedeva, I.A., Tsaplina, I.A., Bogdanova, T.I., Karavaiko, G.I., 1994. 16S ribosomal RNA (rRNA) sequence analysis and phylogenetic position of *Sulfobacillus thermosulfidooxidans*. Syst. Appl. Microbiol. 17, 509–512.

Townes, J.M., Cieslak, P.R., Hatheway, C.L., Solomon, H.M., Holloway, J.T., Baker, M.P., Keller, C.F., McCroskey, L.M., Griffin, P.M., 1996. An outbreak of type A botulism associated with a commercial cheese sauce. Ann. Intern. Med. 125 (7), 558–563.

USDA Nutrient Database, 2016. Seaweed, Irish moss, raw. Available from: https://ndb.nal.usda.gov/ndb/foods/show/3157?manu=&fgcd=

Vanderzant, C., Splittstoesser, D.F., 2001. Compendium of Methods for the Microbiological Examination of Food, fourth ed American Public Health Association, Washington, DC, USA.

Villamiel, M., Del Castillo, M.D., Corzo, N., 2006. Browning reactions. Food Biochem. Food Process., 71–100.

Wang, Z., Cai, R., Yuan, Y., Niu, C., Hu, Z., Yue, T., 2014. An immunomagnetic separation-real-time PCR system for the detection of *Alicyclobacillus acidoterrestris* in fruit products. Int. J. Food Microbiol. 175, 30–35.

Walker, M., Phillips, C.A., 2008. Original article *Alicyclobacillus acidoterrestris*: an increasing threat to the fruit juice industry? Int. J. Food Sci. Technol. 43, 250–260.

Walls, I., Chuyate, R., 1998. *Alicyclobacillus*—historical perspective and preliminary characterization study. Dairy Food Environ. Sanit. 18, 499–503.

Wisotzkey, J.D., Jurtshuk, P., Fox, G.E., Deinhart, G., Poralla, K., 1992. Comparative sequence analyses on the 16S rRNA (rRNA) of *Bacillus acidocaldarius*, *Bacillus acidoterrestris*, and *Bacillus cycloheptanicus* and proposal for creation of a new genus, *Alicyclobacillus* gen. nov. Int. J. Syst. Bacteriol. 42, 263–269.

Wizzard, G., McCook, K., Jackson, J., and Gordon, A., 2002. Quality of bottled coconut water during storage; pH as a reliable indicator of coconut water spoilage pre-processing. In: Caribbean Academy of Science Annual Meeting, University of the West Indies, Mona, Jamaica (Abstract).

World Health Organization, 2004. Guidelines for Drinking-Water Quality: Recommendations, vol. 1. World Health Organization.

World Health Organization, 2006. Guidelines for Drinking-Water Quality (Electronic Resource): Incorporating First Addendum, vol. 1. Recommendations, third ed. WHO, Geneva, Switzerland.

Wrolstad, R., Acree, T., An, H., Decker, E., Penner, M., Reid, D., Schwartz, S., Shoemaker, C., Sporns, P., 2000. Current Protocols in Food Analytical Chemistry (Enzymes). John Wiley & Sons Inc., New York, NY.

Yamazaki, K., Murakami, M., Kawai, Y., Inoue, N., Matsuda, T., 2000. Use of nisin for inhibition of *Alicyclobacillus acidoterrestris* in acidic drinks. Food Microbiol. 17 (3), 315–320.

Yamazaki, K., Teduka, H., Shinano, H., 1996. Isolation and identification of *Alicyclobacillus acidoterrestris* from acidic beverages. Biosci. Biotechnol. Biochem. 60 (3), 543–545.

Yiannas, F., 2008. Food Safety Culture: Creating a Behavior-Based Food Safety Management System. Springer Science & Business Media, Springer, New York.

Yokotsuka, T., 1986. Soy sauce biochemistry. Adv. Food Res. 30, 195–329.

Yong, F.M., Wood, B.J.B., 1974. Microbiology and biochemistry of soy sauce fermentation. Adv. Appl. Microbiol. 17, 157–194.

Yong, J.W., Ge, L., Ng, Y.F., Tan, S.N., 2009. The chemical composition and biological properties of coconut (*Cocos nucifera* L.) water. Molecules 14 (12), 5144–5164.

Zhao, D., Barrientos, J.U., Wang, Q., Markland, S.M., Churey, J.J., Padilla-Zakour, O.I., Worobo, R.W., Kniel, K.E., Moraru, C.I., 2015. Efficient reduction of pathogenic and spoilage microorganisms from apple cider by combining microfiltration with UV treatment. J. Food Prot. 78, 716–722.

Zhang, J., Yue, T., Yuan, Y., 2013. *Alicyclobacillus* contamination in the production line of kiwi products in China. PloS One 8 (7), 67704.

Zierler, B., Siegmund, B., Pfannhauser, W., 2004. Determination of off-flavour compounds in apple juice caused by microorganisms using headspace solid phase microextraction–gas chromatography–mass spectrometry. Anal. Chim. Acta 520 (1), 3–11.

Index

Printed in the United States
By Bookmasters